Aliens

Gold Tenth Planet

MICHAEL

Dedication

To my wife for all her help.

I would like to extend my thanks to all my editors, distributors, and professionals involved in the making of this book. A special thanks to Barry for his cover designs.

A heartfelt thanks goes to my two sons, and a hope that they will forgive me for all my shortcomings. I also hope that anyone else I have wronged will forgive me.

I feel sorry for ALL mankind, but especially to every living thing beneath US!

I dedicate this to scientific discovery.

Contact me and find out more at my website,
www.fortunecity.com/westwood/cerruti/1072

Aliens Gold Tenth Planet
By Mike Brumfield

ISBN # 0-9740390-0-4
Library of Congress Catalog Card Number (Applied For)

Retail $12.95 plus $3.50 shipping. Expect delivery in two to
three weeks.
To order call 615/516-3910 or 631/657-5815.

First Edition, May 2003

ALIENS GOLD TENTH PLANET
Could it be their
Second Coming
2012?

This book is a sequel. It is written as fiction based on real events.

Discover a riveting story of two men whose adventure was made famous in the first book called *The Two Witnesses and The Religion Cover-up*. Their shocking saga continues and climaxes with a struggle for life itself. In the end, it challenges everyone's open-mindedness and willingness to consider new scientific evidence. Will religious people accept these findings when shown physical proof?

These two men embarked on a journey of worldwide discovery, seeking knowledge about man's origins and his religious beginnings. Religious life or death doctrines and ignorance of the world around them drove their search. They did this out of respect for their children's lives. How could anyone make a life or death decision for them without researching everything? They couldn't!

Their search begins and shocks everyone, even themselves.

It led to discoveries of primitive man mining gold for gods/God and ancient alien artworks. This was puzzling. It presented two ultimate questions: Why was primitive man mining gold for gods/God? And could these aliens be the gods of primitive man?

These two questions became their obsession. Please remember these. We will soon be asking you these questions. But first, why did these gods/God need gold anyway? It was obvious primitive man didn't need gold and yet, it became the monetary foundation for all peoples of the earth. Hence, we have the Biblical quote that almost everyone knows: "Money is the root of all evil." Could gold have a universal scientific purpose? Is it possible that these gods/God are flesh and blood, like us? Is it possible that they can't blink and think creation?

They asked everyone. Most answered that God could do anything, like a genie. But they couldn't explain the obvious contradictions. Why did it take six days to create man, three days to raise the dead, and why does he create children to kill them? These are questions no one would touch. There are even "chosen" children, and he sacrifices his only "begotten" one. It's insane to not question this, especially since he foresaw all this. What loving father would do this? We would jail him for life or impose the death sentence. This is a worldwide epidemic.

The Immaculate Conception and word "begotten" sounded like the alien abduction phenomenon.

These two reached a very different conclusion about religion; one that was supported by scientific evidence and logically answered the mystery of God/gods. According to the Bible, the mystery is "finished" with the prophecy of Two Witnesses in Revelations. They are killed for torturing the world with their answer. This really surprised us because it came true. The first book has this title, for that reason. Our theory tortured close-minded religious people. We were attacked, but not killed. You will have to read further to find out if that happens. If it does, we took comfort in these facts.

The prophecy of events read as follows: We are at global war. The mystery of life is solved. The Two Witnesses are killed. But they are resurrected! Then there is contact. It is made by Michael the Archangel to save the earth from destruction. The kings of the earth orchestrate this destruction! We are facing a global nuclear dilemma. Scary, huh? This all climaxes at the same time.

If our mystery is ongoing, this is a logical sequence of events. Our research concludes that mankind is evil, eternal, and addicted to outward beauty (we are them, the one-third imprisoned). We are at the point of global war and threatening the earth's destruction. There is no religious, magical genie God to blink it away. At least our ex-religions haven't produced it, anyway. And the world is begging for one to do the same. This tortures the majority of the religious world. They *might* (you have to finish reading to the end—HOLD ON!) kill us. It would take a resurrection to prove our theory. We can't perform any magic; that's religion's claim. And then finally, global contact would give us peace. At least for the ones who have gotten over their addiction to beauty of the flesh. We are told to do nothing as they save our planet.

The mystery ends in a peaceful protest. And now, a test to challenge your open-mindedness. Then, the hard evidence of the alien existence by primitive man!

Remember the two ultimate questions? Please answer them for yourself to see if you could possibly be religiously brainwashed. If you find yourself answering these two questions with an emphatic and resounding "No," then you could be religiously brainwashed, like we were and the majority of mankind is. To see if this is possible, ask yourself if others can be religiously brainwashed. If your answer is yes, then ask yourself if you could be. If your answer is no, then you are exhibiting the second symptom. Denial. If you get angry and refuse to acknowledge this reality, you are exhibiting the third symptom. Finally, if you are religious and still think you're right, you could be religiously brainwashed.

We were! We were loyal to different Christian denominations that think they're right. Neither has raised the dead to prove it! Uneducated parents following the same family traditions taught it to us. But we soon found that our self-righteousness was a contradiction to "Judge not, lest you be judged."

So we chose to be discoverers, instead. We humbly forged ahead. Tradition was an impediment to our learning process. Since then, we have discovered that today's science sheds new light on the origins of gold and the gods/God image. The Biblical scripture supports our scientific evidence of gold, "Heaven's streets are paved with gold." Gold protects our astronauts from dangerous ultraviolet rays. It is essential in the pursuit of space exploration. Furthermore, it can repair and replace ozone layers. This is documented by the ancient Sumerians. They gave us the word "Jew." It is a Sumerian priest. Their history reveals a tenth planet in our solar system, which had already destroyed their ozone. These gods needed the abundant gold of the earth and created man for this purpose. He was created on the sixth day in a scientific process from the earth itself. Not instantly, like a magical genie would do. Please look at the matching religious and scientific symbols below.

His purpose is also supported by another Biblical scripture; "He was a tiller of the ground." Tilling is physically hard! No wonder they created us. The ultimate mystery to be solved is the gods' disappearance, and what they look like. The gold halo symbolizes the ozone and sheds new light on worldwide alien art. It also gives us a logical outward ugliness—being the reason man became God's image because of our desire for power and love of beauty. We obviously didn't want an ugly little god. This theory is supported by one of many statues in this book.

The Cernes Giant! This also confirms Genesis' "Gods mating with daughters of men because they were pretty, begatting giants". (See front cover of *National Geographic* statue.)

All cultures began with gods/angels in a golden age of immortality. And then they made man! Their fall results from a power struggle. Our existence threatens our planet because of a power struggle! Art imitates life. We are their artwork. A beautiful, competitive worker!

And now, the hard evidence by primitive man! But first, a key piece of recent evidence from our own government. It will confirm the authenticity of primitive man and vice versa.

A flying saucer crashes at Roswell, New Mexico in 1947. This is where they built the first nuclear bomb. It was ongoing. The local paper reported this and the government released that exact same caption on the front page. Anyone can verify this by purchasing a copy. They sell you two. The first report was on July 8 and the second on July 9. The first is small, the second is huge! The second is a retraction of the first and claims it was a weather balloon. These facts remain. There are witnesses. Three child caskets were ordered from the local funeral home. And we now have videotape showing a dead alien body recovered from the crash! It is still not proven to be fake. The Roswell

Alien Autopsy clearly shows a big headed alien with six fingers and six toes. Six!

It supports all the artworks of primitive man's god, including the Hopi art of a big headed god in Arizona. The sixth being appears to be saving the earth and recycling the five obvious figures of man. To the big headed figures, of the Owl Man in South America and Wandjina of Australia. Not to mention the big heads of Easter Island with alien eyes and flying saucers on top of their heads. They are all alike. One! Then there is the head molding of early man. They were trying to look like the gods. And we have an "alleged" alien skull from Nazca, Peru. It was found near the ancient statue of the Nazca god that was called the Owl Man. Primitive man even universally named the owl "wisest" bird because its head and eyes resembled the gods. BIG!

Could mankind be a necessary evil of the universe for the mining of gold? Is it possible that we are them, and have gotten addicted to the power of mankind's outward beauty? This is suggested in the book of Jude, "The angels/gods which left their first estate for a strange flesh." Could our futuristic concept of creating bodies be unfolding before our very eyes? Beautiful workers, spare body parts, and ultimately youthful immortality—is it science fiction? Could all this have happened before? I present it to you, this evidence and more that seems to support this reality.

Could mankind be the machinery for time travel, which ultimately threatens all life on this planet? Could we have already done this on Mars? It is the sixth planet from the tenth, and there is evidence to support this claim. Like the face on Mars. (See pictures.) I have included new ancient evidence to support this existence of this face. The matching satellites (on the cover) and religious symbols alone give us plenty of evidence to confirm this. We can't see these religious symbols, which do match our scientific creations, without an electron microscope.

And finally, could the aliens return in 2012? This is the Mayan end of time and a new golden age. The Hopi Prophecy, called The Great Purification, clearly supports this outcome, as does the prophecy of Michael, in Daniel and Revelations. Challenge everything.

KNOWLEDGE IS POWER, LOVE IS
FORGIVENESS, REVENGE IS EVIL. And
PEACE IS THE ONLY ANSWER. WWYD?

Love, Michael

The Second Coming, 2012?
The Past
January 2, 2001

"The Two Witnesses and Christianity!" I yelled as my eyes flew open. Much to my horror I was staring at those headlines in a newspaper that hadn't even come out yet. I had it in a death grip and was shaking it just inches from my face. This can't be happening! I don't remember having this newspaper. Did someone leave it on the bed as a cruel joke? But how could they? It was tomorrow's paper. And there was that date again: 1/2/2001. I don't remember it being on the bed when I climbed on to do the mystery.

What time was it? How long had I been doing the mystery? It seemed like only minutes. And most of all, I couldn't believe this all has happened just like it did in my dream. Could I really have a best-selling book? I suddenly remembered the dream. I had just experienced it again as I slipped into the mystery. I tried to remember if I had finished it. I never had this dream twice, just once in 1989. That was terrible enough. But I just experienced it again. I know I did! Why else would I be like this?

Sweat was streaming down my face. My whole body was tight with fear. I kept my grip on the paper and slowly looked down to see if I had been shot. I felt pain in my chest but no wounds were there.

What is going on? This paper can't be real. I had to see what time it was. Maybe I had done the mystery all night like I planned. That's it! It must be morning and the bellboy brought me the paper. But, wait a minute. How did he get in? He could've had a key. What am I thinking? All I need to do is look at the clock. That would clear up everything. I turned to look at the clock on the nightstand.

"No!" I screamed in disbelief. Not only was

it just after midnight, but there was my book, *The Two Witnesses and The Religion Cover-up*!

Again, I shook the paper. ""This wasn't here when I got here! I hadn't been doing the mystery longer than a few minutes. Could the aliens have done this?"

I looked around the room; it was empty. I dropped the paper and began to rub my eyes. I was trying desperately to make sense of this. I took a couple of deep breaths and again opened my eyes to look down at the paper. The date didn't change and neither did the headlines. It was just as shocking now as it was in 1989, and obviously moments ago. I was just having this dream again—or was I? Is this dèjà vu, or what? Could I be dreaming now? What is going on? Is this possibly my proof, instead of having to fulfill the Two Witnesses prophecy? Was the dream going to change and we wouldn't get killed?

Oh God—uh rather, oh Aliens—I hope so. I didn't want my wife and children to see my death on national television. I hadn't ever been able to give them any proof of my faith in this dream, and prove to them that I could come back. Hell, I was still unsure, myself. I couldn't do any magic and my mental telepathy sucked. And now this dream just changed. Or did it? What am I thinking, it didn't change. It just happened exactly as I had written it in my book, the ending anyway. Maybe this was my proof that everything would be okay. But how could I give this to my wife and sons? "I'll call them, that's how." I could sacrifice myself. Now I had the courage. I knew why the aliens didn't just openly contact me. That was in my book! That and my disgust of the Judeo-Christian God sacrificing his son. What kind of a father would do this? We would give that father a death sentence. How could people buy this? Especially when their God can have it any way he wants, even perfect if he wants. Worst yet, he foresaw it! This sickened me. I had just spent six years of listening to Christians, Muslims and Jews making the same compromise for this. God wouldn't want us to be like robots. No, he'd rather kill his children! What a sadistic, cruel story!

I thought about those headlines and realized that, unlike then, I now had something torturous to say to the world—at least, to those who glorify mankind as a creation. And that was clearly the

Judeo-Christian religion. But more importantly, Christianity. They had changed the name "Yeshua" to Jesus, and now it lost its importance.

This makes it easy to understand why we would be hated for his namesake. His name confirms our answer about who we are. We are them! I am Yeshua. "The angels which kept not their first estate, but left it for a strange flesh." The angels are the aliens, and they're outwardly grotesque. That's how Travis Walton described them. He even had a movie made called *Fire in the Sky*. The government couldn't hide his encounter. In fact, they participated at every level to find him. He was abducted and had five eyewitnesses to testify when filing the report. They had him for five days.

We had the Roswell Alien Autopsy, too. Man, is the alien ugly! I know that beauty is relative. But the fact remains, we all see outwardly first. If a baby's ugly, we just don't say it. Nobody says, "Oh, what an ugly baby."

I remembered my experience when I was only six. It happened at night! I looked up and saw it looking at me in the window. I freaked out. It was hideous looking. And if that wasn't bad enough, my parents privately discussed it being a demon. I guess they didn't know what else could look like that. But I heard them and went into a frenzy again. They tried to calm me down, but only made it worse by saying that they exist and it was a fact. But not to worry, all I had to do was to say a prayer to Jehovah. Right! Then they sent me to my room.

Lucky for me that I didn't sleep alone. I slept with my brother Jeff. He was just above me in age. I was the youngest. The last of six. The sixth! Wow, 666! What is, was, and will be. Mankind was created on the sixth day.

I always remembered this scripture about the devil when I got older. It was one of my first questions, about an obvious contradiction in my parent's religion. I started telling little kids the devil didn't really exist. The Bible says he isn't real. Besides, the devil is supposed to be male, but is called a whore! The alien of the Roswell had six fingers and six toes. I thought I knew why. I pointed this out in my book. Robert Urich didn't get it, but neither did anyone else. The scientists and other experts scrutinizing the evidence didn't get it, either. No one did, but me. I was the sixth child; man was created on the sixth day; and we had the nuclear symbol matching the Jewish star! I theorized the angels/aliens created us scientifically for power through beauty of the flesh. And man's condition exemplifies the power struggle of the angel story. The past, present, and future all wrapped up in man! We had all the evidence of a scientific creation! DNA matched the medical symbol, and the yin and yang matched the sperm and egg.

We had reached this conclusion about mankind being the one-third of the angels cast out of heaven/the universe, together. We proved the universe is heaven and we are imprisoned in it. That's what tortured religion. They make it a spirit realm. The whole body of the Gospel and Yeshua's teachings is about this question, Where is heaven? That's the religion cover-up! Primitive man knew where this was and even gave us the universal gesture of looking up. They knew it was the Universe! Yeshua said heaven is within us. The universe is made of atoms. This mystery arose in the absence of Aliens/God/gods/angels. But now science confirms Yeshua's answer, and we are on the verge of dematerialization, teleportation and time travel itself! These are all logical events that happen simultaneously with the Two Witnesses solving life's mystery.

Everything in the universe is made of atoms. The conquest of this invisible creator not only answers the question, but also provides us the reason contact will be made. It happens to save the earth from nuclear destruction, according to the Jewish story. The Jewish star is the sign. It matches the atom.

Michael leads the battle with all the angels. My name is Michael! It is also confirmed by the Hopi Purification petroglyph in my book. And it all happens because atoms don't have an ending or beginning; they just change shape. The invisible God of the Jews matches the scientific definition of the atom. It creates through natural selection/ evolution to achieve perfection and multiply. The angels are its perfect creation. They are the gods of primitive man, their creation. And we are something that can never maintain perfection, their addiction. Mankind is the disease of the universe. We fall prey to outward beauty and power. We don't like being equal and especially don't like being ugly.

I was the one with the alien evidence. Jake never went there. He only considered my answer about the alien creation when I made a connection to his Moth Man's and the alien's large eyes being the same. He finally agreed to the plausibility of my theory after I showed him the ancients had scientific symbols of creation. They had DNA!

He had a hard time with anything that denied his ego. And even he questioned my theory that the Jewish star was the atom and reason for contact. That's the reason the book is set in New York City, well, that and the dream. It dictated that! I predicted that the date of the Mayas would correspond to a world event. A new beginning in New York City, the world's largest city, and in the leading nation of the world. Coincidence? Little did I know that New York would win the Olympic bid for 2012! It would happen one year from now!

This also meant that I predicted this moment, or at least, my dream did. I suddenly realized I must have been dreaming now. It couldn't have changed. It was now famous worldwide in my best-selling book, *The Two Witnesses and The Religion Cover-up*! But what was about to happen? I ended it with the *Aliens Gold Tenth Planet*.

I realized that I could change the dream. What am I thinking? It has changed! I could call Angel! I hadn't done this in my book. I suddenly began to cry uncontrollably. It felt as if the weight of the world had been lifted off my chest. I forgot to call last night, just like the dream, and yet it didn't happen. I woke up! Or did I? I was so afraid that the dream would happen, even though I knew it wouldn't be my death.

Jake wasn't here. Hell, I didn't even know if Jake was the other witness. And I only had this dream for proof of being one, myself. But was I? And now this! This dream isn't the same!

I had to look at the clock again just to make sure that I wasn't dreaming now. The dream only showed me getting shot at. Wait a minute—the newspaper! If it *is* tomorrow's, I mean today's, I laughed as I remembered the time. It was after midnight. Then the article would tell if this dream happened. I picked up the paper and glanced at the clock as I started reading.

"No!" I screamed. It displayed the time, 6: 59! It was only seconds before my wake-up call. I immediately looked back at the paper. I was terrified of the outcome. There it was, I *do* get shot. "Oh my God!" I screamed in horror. "This dream is going to happen!"

I desperately read on. "Oh, thank God! What am I saying—thank you, Aliens. I live!"

I dropped the paper and breathed a sigh of relief. It took a few minutes to gather myself. I had to call Angel. As I looked up to get the phone, there standing before me are three figures. It stunned me. Clearly, there are two aliens, one on either side of a tall, thin, naked man standing at the foot of the bed. The aliens from Roswell, New Mexico.

The aliens? I freeze. I panic. I am speechless. I can't say anything. I just look. They look at me and I look at them. Even through all my fear, I feel at peace. Then, just as suddenly as they have appeared, they are gone.

I dropped the paper. I am just blown away. What is going on?

I frantically look around. I can't believe my surroundings—a motel room in New York City. I picked up the paper to get some facts. I look first at the date and see those horrifying numbers: January 2, 2001. This couldn't be real. 2001? This is 1989. The last time I could remember I had been in Ohio. I live in Ohio. What am I doing in New York?

It's 1989! This means I'm 12 years into the future. As I start reading the headlines, I can't believe my eyes. "Showdown! The Two Witnesses and Christianity." As I begin to read I am startled to be reading about myself. Apparently, I am one of the Two Witnesses. Who is the other witness? As I read, I can't believe it. It is my best friend Jake, from Nashville. Is he still in Nashville? I try so hard, but I can't remember anything. I read on. My Uncle Brice has become famous for his carvings? The story reads like a Hollywood movie script: I have invented a recycling trashcan; I have sued the Jehovah's Witnesses and won a lawsuit against them; Jake and I have a hit song; and biggest of all, there it is—the Two Witnesses. The Two Witnesses of the best selling book! Jake and I are the Two Witnesses who are to battle it out with Christianity! What! What is going on?

Without warning, the alarm clock radio

goes off and starts playing our song. Not just any song! They are playing our hit song, "One of Us."

This is all too unreal and unbelievable. I sit back and close my eyes, trying desperately to absorb all this, to remember something, but I can't remember anything. Just as soon as I close my eyes the phone rings. Naturally, it must have been the phone, bang, bang, bang—I must have been dreaming, and the phone rang.

It felt like an eternity, but it could have only been moments at best. The phone rings again. I finally pick it up. I expect the caller to be Jake. I just know this has to be one of Jake's bad jokes, but it isn't. It is the lady from the front desk, and she has these words to say: "Hello, Michael. It's your wake-up call."

It is my wake-up call, all right. I need to wake up! I hang up the phone. I know this has to be a dream. I am in the middle of a nightmare. This cannot be real. This is 1989! I am in Ohio. Jake is in Nashville. I don't play any music. My uncle isn't famous. None of this is possible.

I cannot remember anything. And most of all, this Two Witnesses thing! What are the Two

Witnesses? I search frantically for a Bible. I am in a motel room, after all. There has to be a Bible around here. As I get the Bible out of the nightstand, I turn and look at the paper again to see where the scripture is. It is a Revelations Prophecy, Chapter 11, and The Two Witnesses. I turn to it, and as I read, I can't believe my eyes. The Two Witnesses are killed! Killed, for torturing the world with their prophecy! What could I have to say to the world that would torture it? I didn't know anything. I am confused. I am shaking with fear. I read on. Wow, a second resurrection! "Thank God!" I scream in utter relief.

But how could we be the Two Witnesses? Jake and I! Jake? I immediately grab the phone to call Jake. I need to make some sense of what is going on.

But just as I start dialing, there is an urgent pounding on the door. "Michael!" They are hollering. "Michael, you're going to be late! Let's go! Come on! The limo is waiting! Let us in!"

Let them in? Who are they? What is going on? I can't let them in. I am panic stricken. Fear is gripping me so tight that I can't move, and yet

I have to move. Again I try dialing Jake, and they pound more frantically as I dial. Time is running out. I glance at the paper one more time, wanting desperately for this to be joke. It looks real. The dates don't lie; the headlines don't change. Then I see it under the paper—my book! Could it be? It is! *The Two Witnesses*. Obviously, Jake and I are the Two Witnesses! And yet, I am alone. I am suddenly more afraid than ever. This has to be a dream. I have to talk with Jake.

Just as I finish dialing, I hear the mob outside calling, "Michael, Michael! You're going to be late!"

Then I hear it. They are going to open up the door. I hold the receiver to my ear and listen as the connection goes through. I hear the key go into the door and it is starting to happen. As the door opens, Jake's phone is suddenly answered—by my wife! At least, I think it is my wife. And it is. But not my wife from Ohio. It is my wife from Nashville. I have married a woman from Nashville? I am petrified. I can't make heads or tails of it.

As they come across the room, I try desperately to talk to the woman. The woman is crying. She pleads, "Mike why didn't you call? Why didn't you call last night?"

I cannot get her to understand I don't know who she is. "I need to talk with Jake. I have to make sense out of this. Please, please, let me talk with Jake!"

They are closing in on me with movie cameras and lights, and the room is filling with people. I am losing control of the situation. With phone in one hand and the book in the other, I am being dragged out. They are tearing at me as I cling desperately to everything. They manage to get the book away from me and much to my disbelief, toss it into a trash can—not just any trash can, but a Michael Recycle Center. My trash can! I can't believe it. This is all happening way too fast.

I scream into the phone, "Jake! Jake!"

Suddenly I look up in horror to find a preacher running at me with a pistol in his hand. Bang, bang, and bang!

I don't wake up this time. I *was* awake. The dream came true. This was real. I felt the impact like someone hitting me in the chest with a sledgehammer. Then another one, and another until I went down.

I thought about the phone call to my wife. Did she call me last night? That had to be what woke me. Here I was dying, and I was worrying about my family.

What am I thinking? Wait a minute. I'm not afraid. I'm alive. Well, at least I think so. I suddenly remember that I was talking to Jake on the phone. But I can't feel the phone in my hand. I try to communicate with the reporter who is above me, pumping on my chest. Pumping on my chest.

I screamed hysterically. But no one heard me. And then I knew! He said my eyes were closed and that I had no pulse. What!

"I'm alive!" I screamed. He didn't hear me. I screamed it again, and then suddenly I was moving. He didn't hear or see me. Hell, I couldn't see me. But I could see him. I was above him! I knew then. I was having an out-of-body experience, and still moving. I knew where I was going.

Was I ready for this? I scared the hell out of myself once before, when I was six. I'm sure my ugliness hasn't changed. I was older, though.

And I had seen the Alien Autopsy. It wasn't that bad. Would I be just as scared now? This wasn't television or research; it was real. I just knew that I was going back to myself, the alien. I continued to move out of the room and accelerate. Or did I leave the room? By now things were a blur.

Then instantly, I was in the black tunnel that everyone talks about. I was glad. I got nauseated at fast movement. They call that vertigo, I laughed. Wow, vertical! Just like this OBE. I loved learning. Listen to me. Here I was dead, or at least, dying, and I was still laughing at life's word game. Man, it is complex. And we only have a short time to get it, yet we keep making new words. "Sinonyms" is how it should be spelled, because that's the sin. Words became plentiful and confusing. I guess that's why we say it best when we say nothing at all. Like the mystery and the famous country star Keith Whitley's song. Listen to me! I sounded like a country music star wannabe. What am I saying? I am a country music star. And now I needed to shut up.

People were talking to me. People in the tunnel! I started seeing people I knew, that had died! This was real and they were confirming it.

Then suddenly I saw him. I couldn't believe my eyes. It was my dad. I'm getting a chance to see him again.

"Dad," I cried out as he came toward me. We hugged, what seemed like forever. It was so wonderful. There was so much that I wanted to ask and say. But just as we separated and I started to ask him something. I was moving again. Had I even stopped? Could he see me, my body? Did I have a body? I didn't know. I still couldn't see myself, but we hugged. I must have a body. I had so much to ask, but I was moving again and couldn't stop and talk with my dad.

He just kept smiling and saying that I was right. "Right about what, Dad?" I pleaded. "I don't care about right or wrong, Dad. I need to know if you can see me, please," I begged mercifully.

"Right about who we are, son," he said, smiling as if he didn't hear my last question. "Now go, and don't be afraid of yourself. This is time travel. That's why you are moving and can't stop this now. No one can. We all have to return. And it is the mystery we must all find. Don't be afraid and continue to do nothing. Everything came from nothing, the atom. And when Adam becomes atom, this mystery will finish itself."

"But Dad," I begged and sobbed. "I need you!"

"No son, you found the mystery and that is all you need. Mankind is the mystery. You did the mystery and I didn't. I must learn and do nothing when the time is right. And we both know where time is."

"No, Dad," I begged. "Please don't stay. Please don't stay with mankind." I knew the horror that he could go through! He caused a lot for me. Funny though, I only remembered the good times now. But I understood how easy it was to do this. I had done it to my sons. Life was hard, when you're poor and raising a family. Hell, I couldn't compare to my dad. He raised eight and lost three. I only raised two, and had royalties from my wife's famous grandfather.

"I'm sorry for ever hurting you, Dad!" I yelled. "I love you." I was moving faster now, and had my back to where I was going. He yelled back that he loved me, too.

I turned around and was immediately

blinded by the bright light. I was moving toward it faster and faster, when suddenly I saw it. It was an alien sitting Indian style and making circles with each thumb and index finger. I was doing the mystery. I started moving toward it so fast that I couldn't see anything else. Where was that bright light coming from? I couldn't tell, except that I was definitely in a black tunnel. The light was everywhere, but it was brightest around the alien. And then, just as I was about to collide with the alien, it opened its eyes. I flew into them! They were so black. But it wasn't black. I was immersed in the light, still moving. I was the light. Huh, and the tunnel. Was this the fiber optics of the universe, we are atoms and it is all one? We can experience it all. The question is what body we live in. This all reminded me of warp speed in *Star Trek*. I realized that the light was coming from the stars of space. I was moving so fast. It all just seemed like one light. Then colors! Yes, what a relief! I loved color. The white was now a glowing red. Man, what a panoramic, brilliant blue. Wow, now green.

Suddenly, I was back above myself, but still moving toward my body. The room was full of panic, as they continued to do CPR. Then I gasped as my body jerked and my eyes opened. I zoomed into the pupil. I was the pupil! I was alive! I grabbed the reporter's arm. "The pupil," I cried. "Is like the tunnel. The light is in the eye. The light surrounds it. It's fiber optics, man. Universal fiber optics. The universe is a living communication system, everywhere. We are the pupils!" I cried. "Help me, please?" I tried to pull myself up by his arm. "I've got to call my wife!"

I knew she would see this. What am I thinking, she had to be watching. The whole world was watching! The reporter just sat there on his knees in disbelief. He was in shock. I saw his mike and immediately grabbed it. I let go of his arm and struggled to sit up. The cameraman immediately lifted the camera to his shoulder and started filming again.

"Angel, Little Jake, Jimmy, I'm all right!" I cried. "Please don't worry." Suddenly the paramedics were there and putting me on a stretcher. They reached for the mike I was holding. I was clutching it for dear life. "Wait," I cried. "I have to tell my family that I'm not going to die."

I was looking for the phone frantically.

"You just did, Michael. Now we have to go if you want us to save your life."

I wasn't going to let go of that mike. If I couldn't get the phone back, it was my only chance to comfort my family.

He gave up, finally, as they picked me up and rushed me down the hall. I immediately started telling Angel and the boys that I had an out-of-body experience. I was crying and going on and on about how beautiful and exhilarating it was. "The mystery is like fiber optics! Universal fiber optics," I cried, and kept repeating. "I saw myself, and I was the alien. And I was doing mystery! I was the light. I was making it, and I was it. It must be the way. I remember that much, and then I started losing consciousness again. I didn't remember anything else. I didn't remember being the alien. That's what they wanted. That's what I wanted."

I wanted it all. All the answers. "I want to know how we travel from this body and back, voluntarily." Was it through the mystery? I had to know. Was it spontaneous combustion through some kind of nuclear chain reaction? What was the science behind it? Was it instantaneous because light is so fast that you can't see it travel? It obviously seemed so. I knew one thing—I wanted to be the alien. I wanted to know what I used to know. "I want to know what I used to know, for my boys. Please," I begged and pleaded.

Suddenly, I felt someone shaking me gently. "Michael, you're all right."

"No, I'm not. I want to experience knowing everything," I mumbled. "I want to stop the hell, the pain, for my boys, please."

She was telling me the boys are all right. But I kept explaining. I thought I was in front of the camera. I finally realized that I was in the recovery room of a hospital.

"You're in a hospital Michael. Your family is fine. Please calm down, please, Michael. You're going to hurt yourself worse. Please calm down."

I could tell she was real. This did happen. And she was serious about me hurting myself more. I was! I started coughing. The pain was unbearable. I blacked out. I woke to her telling me what happened. She was stoking my cheek and crying.

"It's all right, now, please don't cry. I'm okay. What happened? What day is this?"

She broke down and starting telling me that I died.

"I know," I replied quickly. "But what day is it?" I asked again, immediately. "What day is it? I need to know, please. Just tell me."

It was dark and I knew that I had to have been unconscious for awhile, at least since this morning. "It is the third, isn't it?" I asked with tears in my own eyes.

She looked at me with a look that was totally wonderful. It was so full of love. It was overwhelming! It was the way my mother first looked at me when I was placed in her arms. She looked that way. She looked that way when the doctor delivered me from her womb and laid me in her arms. She was smiling with tears streaming down her cheeks and said, "It's all right, don't cry. I love you."

I remember now. The nurse was doing the same thing.

"What day is it, and why are you crying?" I asked again. "I need to know.

"It's a miracle," she started telling me. "You were dead for so long. It's impossible. It had to be a miracle. It isn't scientifically possible to be dead for so long."

"How long was I dead?" I asked and then saw her nametag. "How long, Gail?"

"You were out for three hours. Three hours," She said in disbelief. "And that isn't possible. You bled out."

"Then, it is the third, right? I've got to talk to Angel and the boys." I started getting out of the bed.

"Michael," She said softly as she gently stopped me. "It's not the third. You've been here for three days."

"Three days?"

"Yes, three days. And you weren't here for much of that."

"Three days," I remarked in shock at her answer. I fell back onto the bed and just stared off in a trance.

"That isn't all, Michael," she said again softly. "You died ten times."

"No way!" I yelled out. "There's no way!" I thought about the tenth planet. This was bizarre. I didn't believe in miracles. At least, not the religious kind. What am I saying? I did believe anything was possible. That was a paradox. I just didn't believe in God healing and helping winners or killing. Heck, what am I saying—I didn't believe in their God. I believed in the Aliens. They could do things, I suppose, obviously, through advanced technologies that are temporarily beyond our comprehension.

"Yes, you did," she said, suddenly bringing me back. "And not only died ten times, but you were out for a total of twelve hours!"

I just lay there. I couldn't believe my ears. "I've been here three days, died ten times and was out a total of twelve hours! Come on," I said as I suddenly turned toward her.

"Yes, you were," she repeated again. "Three days, ten times, and twelve hours to the second."

"Whew," I just gasped, grabbing my forehead. "I can't believe it."

I know she clutched my hand. "It's unbelievable, isn't it?" She almost squealed out in excitement. "Do you think this is it, Michael?" She asked feverishly. "If it is, I've got to go tell the doctors."

Then the door opened. It was the doctor,

then someone else, a cameraman. Then someone else—yes, yes, it was Angel and the boys! "Angel, thank God!" I couldn't stop saying that.

They all broke past the doctor and cameraman, reaching the bed together. Little Jake jumped and just threw himself into my lap as I hugged Angel and Jimmy. He wasn't so little anymore. He was crying and squeezing my head hard. Angel was trying to kiss me and keep them both from hurting me at the same time.

"Jake!" She cried, grabbing him as he was going for my chest. "You can't hug Daddy s chest."

"The heck he can't," I said, grabbing Little Jake and Jimmy at the same time. I started kissing them as much as I could. Much to my surprise, they let me. Heck, they were even kissing me back. My oldest was really squeezing me. He had stopped the kissing, but not Little Jake. He was pulling my face around to kiss me more and telling me that he knew I wouldn't die.

"You said you would be back Daddy," he just cried. "I knew you would. I told them all, and they wouldn't believe me."

Maybe they shouldn't hug me. I was about

to pass out from the pain. "I wouldn't ever leave you, Little Jake. You know that."

"I know, Daddy," he cried. "I kept telling them you wouldn't die."

I reached to pull loose from Jimmy's grip. He was crying convulsively, saying that he loved me. "I love you too, little buddy. But you've got to let go of my back and chest." I winced with pain. Angel immediately grabbed Little Jake. I slumped back onto the bed as I broke from his grasp. He saw that I was in pain.

"I'm sorry, Dad. I didn't mean to hurt you." He started to cry harder. "No, don't cry, son. You didn't hurt me. I'm all right."

He cried out, "You're bleeding!"

I looked down at the bandage that was wrapped around my chest. It was bloodstained. I started to reach for Jimmy again. I didn't care about my pain. "Come here son, I'm okay."

The nurse immediately intervened and stepped in between us. She hugged Jimmy and then looked intently into his eyes. "Your dad is just fine, but he needs those bandages changed, and some rest. Now, don't worry anymore and please just help your mother. It looks like she's got her hands full."

Little Jake was still struggling to get to me. I think she must've given Jimmy that same look she gave me, because he quit crying and went straight to Little Jake. "Jake," he said in the most stern and yet loving way I've ever heard. "We have to help Mom, now. Dad is going to be all right. Besides, look around. Dad needs protected from all these people. He looked right at the camera crew. He doesn't need to be bothered and we have to do that. We have to be the men of the house for Dad."

Little Jake quit struggling and just puffed out that little chest. "Don't worry Jimmy," he said as he quit struggling and wiped away his tears. Me and you, we'll take care of everything. I promise."

Jimmy hugged him like I've never seen before. "okay little buddy, I couldn't have a better partner."

Then they grabbed and hugged Angel. It was one of the most beautiful moments of my life. I had never seen Jimmy and Angel hug. They both came from non-affectionate families. I sometimes wandered if they ever would hug. Heck, it was only recently that he started hugging me. I was trying to hold back the tears.

The nurse couldn't hold back hers, so she turned and started to work on my bandages. "All right Jimmy, it's time to start work. Your dad needs his rest and I've got a job to do."

"Okay," Jimmy said as he started toward the door along with Angel and Little Jake. "We love you Dad, and don't worry about us," he said without looking back. "We'll take care of everything, won't we, Little Jake?"

"Yeah Dad, you can count on us," he said looking back. Jimmy kept him going toward the door. I knew he couldn't look back and neither could Angel. I was glad they didn't. We all would've bawled.

Besides, Little Jake was ready to take charge. He grabbed the doorknob before Jimmy could get a chance. "Let me talk to them, Jimmy. Grownups are suckers for us little guys."

I just laughed along with the nurse. I guess my show business side had worn off on him.

"Besides, I can't wait to tell the world who I am. I am the son of the most famous man in the world." He looked right into the camera. We all laughed.

"Well, I am too." Jimmy remarked quickly.

"Yeah, but I'm the one who wrote a song with him. And that's what the world knows him for, not the book."

"Well, that's not exactly true," said the cameraman. "We are live and this is playing on every major network in the world, including CNN."

Jimmy started to reply, but Angel broke in. "And your brother also helps with the music, young man. So let's not argue over who is going to do the talking. I'm his wife and it's my job to do this."

Little Jake started to say something again, but I interrupted quickly. "She's right, Little Jake. Now you and Jimmy help guard her from the crowd. Especially you, Jimmy. Jake is still little for his age."

"But I'm almost ten and a half, Dad," he begged.

"I know son, but you just help Jimmy protect Mom. This is her job."

"But, but Dad."

"Jake, those reporters use a lot of big words, and—" I didn't get to finish my sentence; he wasn't about to get embarrassed.

"You're right, Dad," he said quickly. "I just wanted you to know that I'm big enough to handle this, that's all. And now you just called me Jake, so that's good enough for me." He puffed up real big and then ordered Jimmy to go first. "Since you are the biggest," he said as he let go of the doorknob and clutched Angel's hand.

Jimmy just laughed and looked at me. I winked at him and he got the message. "Okay," he said convincingly enough to be believable. But don't forget, we're partners, right?"

"Right," he said as he gave Jimmy a handshake. "Partners; now let's go. Everything's going to be fine, Dad" he said as he shoved Jimmy to open the door.

Boy, I didn't have to worry about him. I had to worry about Jimmy being able to stop him. However, he was headstrong just like his mother. So I knew she would step in, and she did. "Slow down, young man, and let your big brother lead the way. All right, Jimmy. Let's go. Don't worry about us, honey," she said as she followed Jimmy.

"I won't," I quickly responded. But I knew that I couldn't keep from it. "I love you guys, and take good care of Mom."

"We will, Dad," Little Jake replied as he grabbed Jimmy's hand and nudged him to go. I laughed at the sight of them. Jake was in the middle holding on to both for dear life. Jimmy opened up the door and was immediately bombarded by reporters.

I started to panic and say something, but the nurse just gently pushed me back onto the bed. "Let Jimmy handle it," she said with that look again. I did. The door shut behind them quickly.

Jimmy was doing his job. I realized there was a television hanging from the ceiling. I immediately grabbed the remote and turned it on. The nurse started to take the remote from me, but all I had to do was look at her this time. She knew my thoughts. It was weird, but we read each other like a mother and son. I watched intently as she continued to change my bandages. It was my first time to see the actual wounds. There were three holes in my chest and drain tubes everywhere. I just stared in disbelief.

"I know," she said in amazement. "It's a miracle."

"No, it's called a bad aim," and then I laughed. It hurt as I laughed and she scolded me for making fun of what she thought was a miracle.

"You of all people shouldn't make fun of a miracle. You're very blessed."

"Look Gail, I don't believe in miracles, except for life itself. I believe in time travel. That's a science…but if we travel back in time and correct a mistake, does that make it a miracle? No. It makes it a science. To have a life with no mistakes would be a miracle. Why don't we question the logic of this magical genie God who can perform miracles if he wants to, but only does it if you beg him enough? And the bottom line is that plenty of children die in misery while their parents beg their asses off. Then they just compromise by saying God needed him. It's time we get our heads out of the sand and question why they stay away from us. Why don't we acknowledge that the heart is a muscle and has no influence on our thinking? Why don't people get the connection to the atom, which is everything, and their God, which is everything. The reason is because it makes them responsible for life instead of their magical God. No, this isn't a miracle; it's a bad aim. The universe is ruled by chaos. Perfection is a matter of timing. Time travel! Besides, if this was a miracle, then it's a cruel God who gave me one, and doesn't stop a parent who is raping and torturing their children. And that's happening right now as I speak. You know that, Gail. And we even call him a father. Come on! He can go straight to hell! He could've had it perfect if he wanted. But no, he foresaw all this and then chose to kill his children!"

"Michael, please don't say that."

"It's true! Why can't you see it? You don't need to answer that. We're all brainwashed in this religious world and Judaism is the reigning king. The story of the invisible God rules. Even though we are digging up lost cities everywhere. That's why I called my book *The Religion Cover-up*. And the cities they're digging up are older than Judaism. But it doesn't matter to people of this faith. They won't believe anything else. Blind faith is really stubbornness, and that's all. I tell my mom and dad that they wouldn't believe the aliens were the gods of primitive man even if they showed up. I have even put the statues of these aliens in this book and found more to put in the next one. The aborigines are an example I am going to use. I showed it to Mom and Dad and told them their story was forty-thousand years old. They have pictures of their gods, and they look like the aliens. I even showed them one from the Jews and the tablets showing a tenth planet in our solar system. It didn't matter.

"In my book, I acknowledged that the Sumerians who gave us the Jews were the oldest civilization. But they're not. We're discovering lost civilizations everyday. Hell, I discovered the Hindus go back millions of years. Nothing that I showed them changed their mind from a religion that is only a hundred and fifty years old. And it is full of fraud. Hell, I even asked them if they thought I could be Michael the Archangel. They said no! Come on, Gail! If you think you're right, you're religious. If you think other people can be brainwashed into thinking their right, but you can't be, then maybe you are brainwashed, too. Life is a discovery. Blind faith is religious ignorance. Ignorance is a disease perpetuated by our parents and then by us if we don't break tradition. I am a scientist and believe the evidence. We are finding statues in Israel of aliens, and that's a fact. The sad reality is that the world's largest religion is making them demons. That's the final frontier. That's what my next book is about. The Second Coming. What if they're the aliens? I discovered 2012 was the date the Mayans say time will end. I'm using that date in my title.

"I joined the Marine Corps and broke tradition from the J.W.'s. I was in Platoon 2012. I don't believe it was a miracle, but I do believe in time travel. I'm asking that question. Is the Second Coming 2012? I don't know, but it's a pretty strong coincidence. And could I be him? I did study to be a lawyer, and Michael's job is to battle all the kings of the earth. And more importantly, could I be battling these religious kings to keep them from nuking each other and destroying all life on this planet? And last but not least, could these gods be the aliens? I have the physical evidence to prove this. It's in my book. And I will put more in this next one.

MICHAEL

"I can't believe I found an alien statue from Israel. I couldn't ask for more, except to have an alien show up. But hey, I am one! We all are! That's my theory! We're all ugly-ass little aliens addicted to beauty of the flesh. That's what gives us power. We all know it. This all happened from power, not love, Gail. If it were a result of love then it wouldn't have happened. And a father would sacrifice himself, not his son. My father would die for me, but would he believe I could be Michael the Archangel? No way! He's religious. That's the final frontier. Science versus religion! And religion won, Gail.

"My dad died last year. But he did finally doubt his religion. He even asked about my aliens. I told him not to worry, that we're all going to make it in time. I gave him my famous quote from Yeshua, 'Many are the first that shall be last, and many are last that shall be first.' That means we all make it, BACK! He even said maybe, when I asked him one last time, if I could be Michael. That was a beautiful moment for him and me. He got some comfort, finally from a scientist. He even learned about the atom. How it never ends. It is timeless, omnipresent, and is everything. Even us. We never end, just get recycled. That is if you don't jump ship first and go back home. Home's where I want to go, Gail. Do you think I could be he?"

Gail was just looking at me with a blank stare. I didn't have to ask her again. The television did it for me. The reporters had the mikes stuck in my wife's face and were asking that very question. They were asking it for the world. She just stood there silent. Gail had turned to watch the television with me. Angel just stared in shock. She didn't know what to say. I don't think she believed it. Hell, I still didn't know if I did. But, I didn't have to worry. Little Jake blurted it out for her.

"Heck yes. My dad is Michael." All of a sudden, Jimmy picked up Little Jake and proclaimed that I was well. Everyone started cheering, but the reporters hollered for them to be quiet. They turned back to Angel and asked her again.

"Well, uh" she stammered, trying to believe it herself. And then something miraculous did happen.

Gail immediately replied at the same time that a woman from the crowd screamed the answer. They said the same thing at exactly the same time.

"Of course he is," she cried out as loud as she could. They all turned toward her as she pushed and shoved through the crowd to get to Angel. Gail put her hand to her mouth and gasped.

"Oh my God," I cried.

Gail immediately asked who she was and how that could've happened. "We said that at the same time. Who is she?" She turned toward me to ask me again. I was crying tears of joy. "You know her, don't you?"

"Yes," I said wiping away my tears. "It's my mother. It's my mother, and her name is Gail."

She erupted into tears as she threw herself onto my lap and hugged me. "You are he, I just know it. I'm sorry," she cried. "I was brainwashed, too."

"Don't worry," I said as I stroked her hair. "We all were."

"Well, I read your book, and I do think you're Michael, the Archangel. After all, you are the one who recognized the Jewish star matching the atom. And nuclear war is the only thing that poses an immediate threat to the entire earth. I don't think that is a coincidence. Besides, your mother thinks so, too." She finished the bandages and was beginning to wash my face. She stopped and started stroking my hair back over my head. I thought about Mom. I had been hit by a car when I was fourteen and spent two weeks in a hospital. My mother had done this every day to take my mind off the pain. It was beautiful then, and it was just as beautiful now. I couldn't get over the coincidence of her name being the same as my mother's.

"Mom," I blurted out. "I forgot all about Mom. Would you do me a favor and go get her, please? Would you please, Gail?"

"Sure I will," she replied, and got up to go get her.

I looked up at the television and saw Mom trying to get past the doctors. They were trying to interview her, but she was desperately pleading with the doctors to see me. "Please hurry, the doctors won't let her in."

"Don't worry, I'll get her for you," she said as she went out the door. I knew she would, too. Where was Angel? I wondered to myself. She must've gone on. Jimmy was definitely doing his job. Or Little Jake, one, I laughed.

Suddenly, the nurse appeared on the television screen. I saw her talk to the doctors and then give me that look again. She looked right into the camera, smiled and gave me the peace sign. Then she grabbed my mother's hand and led her past the doctors and through the crowd of reporters. I laughed because they all just calmly moved aside. "Now *that's* a parting of the Red Sea, if I ever saw one," I said aloud, and started laughing again.

The door swung open and in came my mother. Nurse Gail was guarding the door. My mother rushed to my bed and threw herself into my lap.

"Thank God you're alive," she cried over and over again.

"No Mom, thank the man's bad aim. And I believe in the aliens, remember?"

She lifted her face and looked into my eyes. She gave me that same look as the nurse. "I do, too," she cried. "And I believe you are Michael, the Archangel."

"Oh Mom," I said as I pulled her up and cried with her. I hugged her like it was the last one we would get. I never wanted to let her go. "Hell Mom, I don't know if I am or not," I laughed, trying to choke back the tears.

She just looked into my eyes and said it again. "You are him, Michael."

"But Mom, I beat my son. I can't be him."

"You made a mistake, son," she said softly. "And you asked him for forgiveness. We all make mistakes. You didn't hurt him intentionally. You broke in a moment of frustration. It wasn't right, but you didn't mean it. And you are willing to accept the same punishment in return. That's a beautiful thing. And you are humble about it. I love you, son, and Jimmy loves you. Now, please forgive yourself."

"I can't, Mom. Not until I beat myself. That's what I've got to do. It's easy to beat something that can't defend itself. I have to do the same to myself. That's what I told my brother Paul before he committed suicide. That's what I've got to do myself." I broke down and cried. She held me as I sobbed uncontrollably. We both just cried. My brother's suicide was hard on us all. Mom began to stroke my hair. It was wonderful, and I quickly fell asleep. I dreamed about that car wreck and how she comforted me the same way then.

And then, I went back further in my childhood. All the way back to that moment. The moment that I saw the alien, when I was six. I started screaming. It was horribly ugly. It was staring at me through the window. And it was dark. I was so afraid of the dark. My parents had scared me with their devils and demons. If they never saw one, they should've just said they didn't know if such a thing existed. Instead they were religiously brainwashed, hook, line and sinker. And I was their bait. That's what I felt like. Like I was about to get eaten. I continued to scream, frozen with fear. My brothers were frantically asking me what was wrong. I couldn't talk. I just screamed. And then my parents rushed into the room. I suddenly stopped screaming and just pointed at the window. For some reason, I was no longer afraid. I heard its message. They all looked toward the window. But, before they could get a chance to see it, it left.

"I'll get that bastard," my dad hollered as he ran out the door.

Mom started stroking my hair and rocking

me back and forth. Then she began asking what I saw.

"Mom, it was a big, black eyed, bald mongoloid. And it scared me because he was so ugly." She was stunned by my answer. I turned to her and looked right into those loving, beautiful eyes. "But then he said he was here for me, Mom. He is me, Mom."

"What?" She asked in shock.

"He's me, Mom. He said I'm from the future. We all are. This is all about loving beauty of the flesh. I'm him from the future. I'm addicted to beauty and power, Mom. It said we all are. Please don't let Daddy hurt him. Please, Mommy."

"Son, you don't know what you're saying. It could've been a demon. Let's pray to Jehovah, and everything will be all right."

"No!" I started screaming. "Don't say that! You're scaring me! You're scaring me," I cried.

She began to stroke my hair as she held me. "I'm sorry," she said softly. "Whatever it was, your dad won't let it hurt you."

"It's not a demon, it's not a demon!" I was crying uncontrollably. "Shhh," she said as she continued to stroke my hair and tell me everything was all right.

And it was. I woke up. She was stroking my hair and rocking me back and forth. She was saying everything is all right.

"Mom," I said with a deep calm in my voice. "I'm okay. I'm not afraid anymore. I know who I am. Who we are," I said as I turned and looked into her eyes.

"I know," she replied with sadness in her eyes. There was something different in her eyes this time. Something that I had seen in my own eyes. It was regret. She had the same regret and shame that I did. We both had hurt our children. I knew she didn't mean to. She was religiously brainwashed. And then I realized that I was, too. I remembered the quote that I heard so often growing up. "Spare the rod and spoil the child."

"It's okay, Mom. I forgive you. You didn't mean to hurt me. Wait a minute. I don't know why I said that."

And then we both looked at each other in disbelief. "Did you dream what I dreamed?" I asked excitedly.

"It wasn't a demon!" She blurted out. "There isn't any such thing. I'm so sorry, I'm so sorry!" She wailed.

"It's okay, Mom. It's okay, we both had the dream. Everything is going to be all right." She kept on crying. I held her face and looked into her eyes. "Mom, if we both had the dream, then maybe we are the Two Witnesses."

She looked surprised. And then she smiled. "Maybe we are, Michael. Maybe we are," she said with a certainty to her voice. "You've made a believer out of me. I still can't get over all the evidence you're finding of these aliens. They must exist."

"They do, Mom, and I think we are them. Why else would they stay away? We're the addicts. We've got to help ourselves and quit seeing beauty outwardly."

Man, I thought to myself. Wouldn't that be great. I couldn't think of anybody more deserving than her. She certainly qualified, in my opinion. She was the best mother a son could have. She sacrificed her own ego for her children. My father didn't. He was unfaithful to the end. And he beat her all the way. She was a beautiful woman, too. She could've had anyone and done anything, but she chose us. What an angel. I was so lucky.

I wasn't so sure about myself, though. I had been a terrible father to my first son. And I hadn't been very good for Little Jake, either. But I was willing to sacrifice myself now. Angel's need for entertainment had turned into an addiction. Actually, we were both addicted from the beginning of our marriage. I tried to cure her, but realized I was making it worse. It was a nightmare for Little Jake. I finally learned to live for Little Jake. And I finally forgave Jake, even though he didn't ask for it. He would never do that. He just said that I didn't really know him. He did that before he left. That's all he ever did. But that was enough. It was a new beginning for us all. Thank God for Mom. There I go again with a Christianity slip. Again, I laughed. Mom asked me what I was laughing about.

"Oh, nothing."

She gave me that look again.

"Well, I just said 'Thank God' to myself. And I just corrected you a moment ago. A moment ago—it's daylight. We must've slept for a while."

And then she looked at me and said, "You've got to go on, Mike. Some people never face up to their mistakes."

I was shocked. She must've read my mind. What am I saying? She has gone through this nightmare with me. "I know," I said and hugged her. "If I want to be forgiven, then I must forgive."

"And you will be," she said, so sure of herself.

"I hope so," I replied weakly.

Then I realized we all will be. We both looked at each other.

"Yes," she said calmly. "You wrote it in your first book, and you will put it on the cover of this book."

Then we both said it at the same time. "Many are first that are last and many are last that shall be first."

We hugged again. I loved her so much. That's the hardest thing to do in life, to love even those who hate you. This body wants revenge.

The next few weeks flew by. Angel brought the kids every day. Mom never left—she stayed. The hospital made arrangements for her to stay

Wiltshire, England

Nazca, Peru

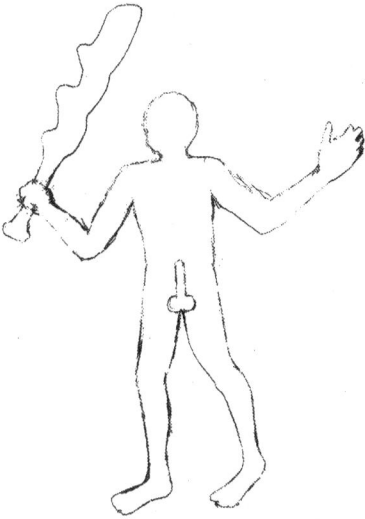

Cerne's Giant

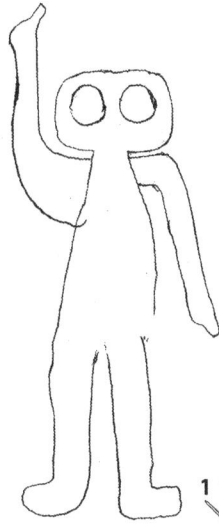

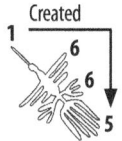

Owl Man

A Hopi Prophecy

"The great purification"

Oraibi, Arizona

The Five Faces of Man

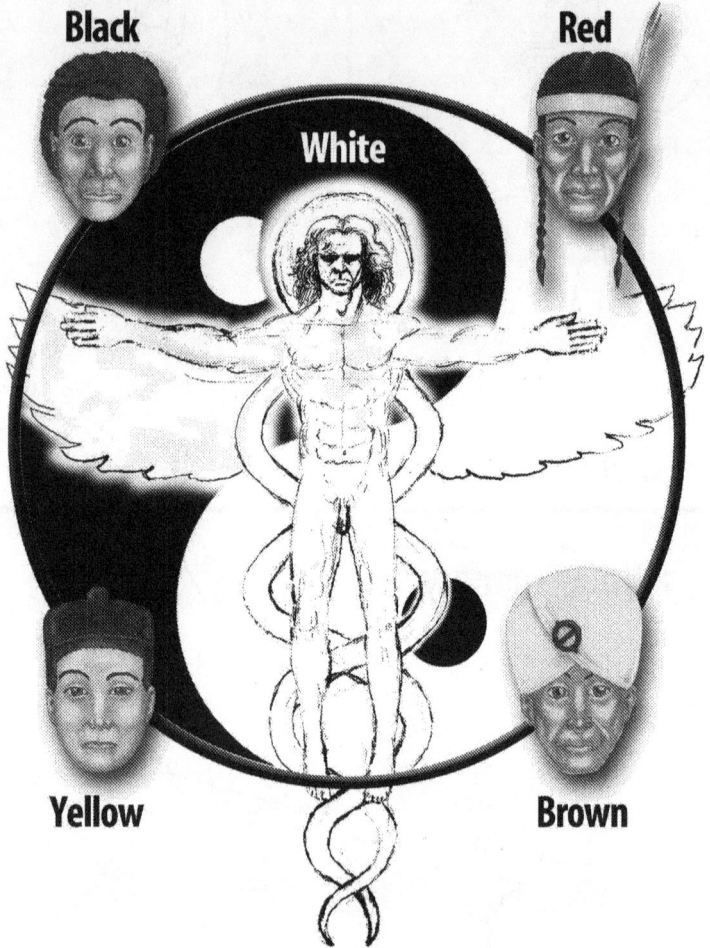

Black

Red

White

Yellow

Brown

1. Why does Christianity make Jesus white?

2. The different colors and facial structures indicate the competition to make the "prettiest" human. This "prettiest" factor is evident in Genesis 6:4 and the fall of the angel story!

The Starchild controversy

SINCE FEBRUARY 1999 *a bizarre looking skull, known as the Starchild skull, has been exhibited at UFO conferences and heavily discussed in UFO journals.*

The Starchild skull is alleged to be the remains of an alien-human hybrid.

Legend of the Star People

According to the Starchild Project, an organization that wants to arrange DNA testing of the skull to prove an incredible origin, the skull was discovered in the mountains of northern Mexico. Indian tribes from the region have legends of Star People – beings from the sky who visit Earth to impregnate local women before returning years later to retrieve the hybrid infants.

Big head

The skull has several strange features that suggest it is not human. It has a massive brain capacity, flattened rear, shallow eye sockets, and is missing the front sinuses.

The Starchild Project claims to have consulted over 50 experts, the vast majority of whom argue that the skull is that of a deformed human child.

Most experts say that the Starchild skull is that of a child suffering from hydrocephaly, a disease in which fluid builds up on the brain and makes the skull swell.

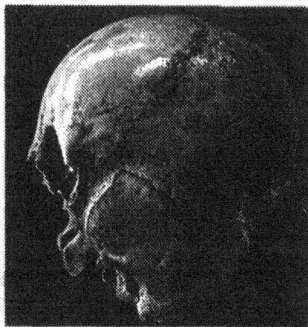

■ *The Starchild skull is far from normal. But is it from an abnormal or cradle-boarded human infant, or perhaps an alien-human hybrid?*

It is also widely thought that the skull has been cradle-boarded. Cradle-boarding is the practice of strapping an infant's head to a board and causes flattening of the back of the skull. It was practiced in the area of Mexico where the skull comes from. The Starchild Project argues that close examination of the skull rules out this explanation, and is attempting to raise funds to pay for DNA testing – the only way to be certain of the skull's origins.

"Beings from the sky" Is this the owl man? This skull supports this. Also "impregnates women" supports the cover art of the mother goddess statue. Evidence of aliens being flesh and blood beings and these "sons of God" in Genesis 6:4. It supports my theory that they are not religious "spirit" magical beings. However, I conclude the atom, which makes everything, is spirit. Read on.

1. Notice alien eyes of Easter Island statues, Jewish statue, then go to (1) and see owl man eyes, covers for the aliens on front and back.

2. The Giants [common to all cultures: see Cernes giant (1)] are a product of the daughters of man breeding for power.

3. Notice the saucer on top of head tells us where they live just like owl man (1) and Starchild legend (3) spaceships. Six rock strands coincidence? Lava rock symbolizes propulsion.

4. See the obvious match of saucer in painting and Easter Island "alleged head dress".

5. Notice ears of single Easter head matching ears of mother goddess statue on front cover art.

Flying saucer

Flying saucer

DNA similarities make world seem smaller

Survey says any two people 99.9 percent identical

By LEE BOWMAN
Scripps Howard News Service

Although everyone's genetic makeup is unique, scientists have found that populations from different parts of the world still share more genetic similarities than had been thought.

The results of a computer analysis of DNA from individuals representing 52 populations around the globe, published today in the journal *Science*, make up the largest such global survey of genetic diversity, and should help studies of ancient human migrations.

Those surveyed were broken into five regions: Africa, Eurasia, East Asia, Oceania and the Americas. Differences among individuals within those groups accounted for 93%-95% of genetic variety, according to the international team led by Marcus Feldman, a professor of humanities and sciences at Stanford University.

Compare the genetics of any two people, and the matchup will be about 99.9% identical. The research team accurately pinpointed the ancestral content of virtually every individual from Africa, East Asia, Oceania and the Americas. ■

(handwritten annotations in margins)

1. How could Hopi medicine man know of five races, let alone the alien god?
See illustration page (1).

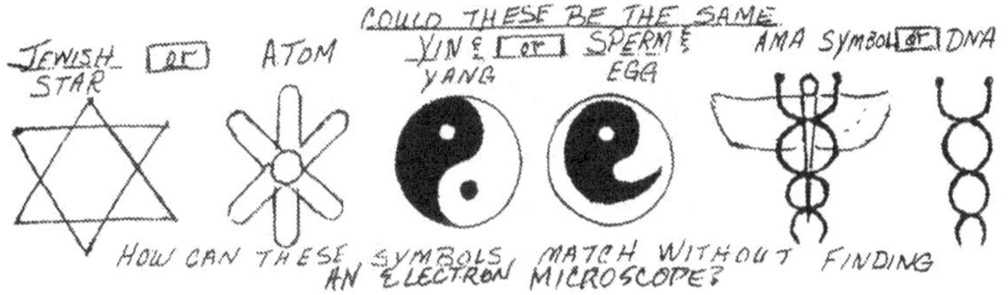

COULD THESE BE THE SAME

JEWISH [or] ATOM   YIN & [or] SPERM &   AMA SYMBOL [or] DNA
STAR   YANG   EGG

HOW CAN THESE SYMBOLS MATCH WITHOUT FINDING
AN ELECTRON MICROSCOPE?

2. Notice alien head on mother goddess statue. This supports what gods look like in Genesis 6:4.

3. Notice asexual organs on alien. This confirms Yeshua's description of angels and explains why they don't give their hand in marriage.

Cave drawing of aboriginal god Nandjina. Notice similarity to owl man (1). Also see halo around above head. This supports 10th planet story of gold replacing ozone and "who" was mining it before they created man as a "tiller of the ground" in Genesis. Man's purpose supports skeletal discoveries in gold mines.

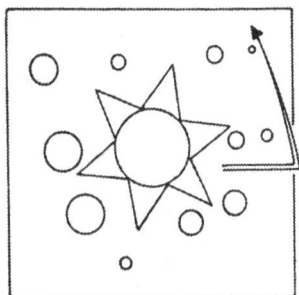

1. Notice cave grid looks like computer grid of space. Also star is sunlight just like the Sumerian clay tablet below. The clay tablet is dated circa 13000 years. The cave drawing is much older.

2. The cave astronaut and gemini-looking capsule are also ancient. This is proof they existed before and supports my conclusion. Read on!

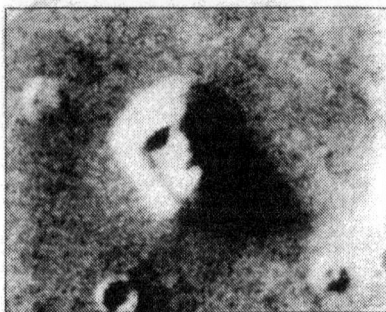

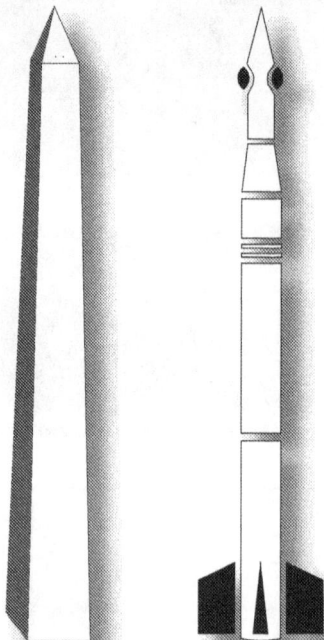

1. Notice the satellite on the clay tablet going from Earth (7th planet) to Mars (6th planet). It looks just like ones today. This tablet is also circa 13000 years. Also notice symbol for Mars matches atom and jewish star. Is this proof that Mars could have had Inca there first and we destroyed it with nuclear weapons? The "man" on mars is in a suit.

2. See how the helmet of Mars "man" matches our pictures of face on Mars.

3. Notice ancient satellite looks like alien head and eyes of nuclear missile. Egyptian obelisk matches nuclear missile. Egyptians called obelisks shem "rocketship".

Exhibit a

Finally my ex-religion's fraud!

HERE is the proof of Fraud. The Jehovah Witness acknowledge Yahwah & Jehovah as the same but don't call themselves Yahweh witnesses. Most importantly they don't recognize that Jeshua is the correct Translation from Yeshua. It is in the book of EZRA 3rd chapter. It is from Yeshua. Why isn't it the same in the N.T. It is easy to prove that Jesus, Jehovah, James, and Christ are all fraudulent, added words. (Why don't all have Hebrew words have two english equivalents?) We even have the correct names in the bible, that is if you accept the changing of Y's to J's. In this case with the Jehovah's witnesses, they actually choose Jehovah over Yahwah! The Four correct translations that are in the bible are Jeshua, Yahwah & Jacob, a Messiah. Please hold them accountable for accuracy. This is not interpretation. Sincerely, Mike Brumfield

PROOF !!!

Whose

24. (a) Yet, should we use God's name, even though we may not be saying it exactly the way it was originally pronounced? Well, we use the names of other persons in the Bible, even though we do not say them in the way the names were pronounced in the original Hebrew. For example, Jesus' name is pronounced "Yeh'su" in Hebrew. Likewise, it is proper to use God's name, which is revealed in the Bible, whether we pronounce it "Yah'weh," "Jehovah," or in some other way common in our language. What is wrong is to fail to use that name. Why? Because those who do not use it could not be identified with the ones whom God takes out to be "a people for his name." (Acts 15:14) We should not only know God's name but praise it before others, as Jesus did when on earth.—Matthew 6:9; John 17:6, 26.

A GOD OF PURPOSE

25. Although it may be hard for our minds to understand, Jehovah never had a beginning and will never have an end. He is the "King of eternity." (Psalm 90:2; 1 Timothy 1:17) Before he began to create, Jehovah was all alone in universal space. Yet he could not have been lonesome, for he is complete in himself and lacks nothing. It was love that moved him to begin to create, to give life to others to enjoy. God's first creations were spirit persons like himself. He had a great organization of heavenly sons even before the earth was prepared for humans. Jehovah purposed for them to find great delight in life and in the service he gave them to do.—Job 38:4, 7.

26. When the earth was prepared, Jehovah placed a couple, Adam and Eve, in a part of the earth already made into a paradise. It was his purpose that they have children who would obey and worship him, and who would extend that paradise all over the earth. (Genesis 1:27, 28) As we have learned, however, that grand purpose was interfered with. Adam and Eve chose to disobey God, and his purpose has not been fulfilled. But

Chapter 24

Chapter 25

24. (a) To be consistent, why is it proper that we use God's name? (b) In view of Acts 15:14, why is it important to use God's name?
25. (a) What things about God may it be hard for us to understand? (b) What moved Jehovah to begin creating?
26. Why can we be certain that God's purpose for the earth will be fulfilled?

---

1. "Where is God?" is the $64,000 question. I thought God is omnipresent! That means everywhere like the atom!

2. They say he's not lonesome but he's all alone. Then they say he creates for others. That's loneliness.

3. Finally they say he creates a heavenly organization of "spirit" sons like "himself". Why not daughters? And now they skip the fall of the angels story.

4. Last but not least. This is the proof that "religion". At least the Jews make these angels and god "spirit" not flesh and blood. This is the ultimate coverup. What if they come back and are the aliens? WWYD?

with me. I'm sure Gail pulled some strings. Gail and Mom became my bodyguards. The media camped outside the hospital. The world wanted me to come forward. Jake wouldn't. Besides, I wrote the book. Jake only did one interview. He said he wasn't the one who wrote the book. That's all he would say. He did tell them that I would give the shirt off my back to anyone. That helped the world's perception of me.

Jake told them that I wasn't trying to destroy church, just unlock it and give to people. Stop making life a right or wrong issue. That was enough, though. Christianity says they're right, and they weren't backing down. They came to me. I mean they *all* came. The parking lot was filled with vans from every country around the world. Jimmy told me that he had everything arranged for my first interview. I have to admit that he took care of every detail right down to my trip home. Of course, that would only happen with my two bodyguards. I laughed. That day finally came. I thought I would be prepared, but I wasn't. I had already been singing in front of huge crowds. However, they loved me. This was different. Very different! I had never answered questions about my book. Americans were angry at my anti-patriotic remarks. The other countries were trying to capitalize on it. It was a media circus.

Jimmy had set up a pressroom in the hospital cafeteria. I apologized immediately to all Americans. I explained that patriotism keeps the world divided. That was the reason for my negative remarks. I went further and explained my stance on religion. I was a scientist and had written this book in hopes of sharing my own experience with religious intolerance. It's a fact of religion. It's not my opinion. I was about to prove it to the world. The world is religiously brainwashed.

One of the reporters challenged me to do a worldwide book signing tour to let the people judge for themselves. I told them I would. The crowd cheered. I was stunned. Nobody booed.

Of course they wouldn't. They were on TV. That was the world's police. When knowledge increaseth was now. We were a global world. WWW rules! I told them that I would do every signing live. I knew that wouldn't stop them from attacking me, but it would reduce the number of attacks. Hell, we had Saddam Hussein and North Korea's president. Both were cruel leaders, even with television. But, time and knowledge would win. I knew that. It already was.

I just hope this Two Witnesses thing happens quickly. If it's me, that is. I had seen a lot of suffering already. Deep down, I felt it was me. That was scary enough. Then there was the alien evidence. My evidence, not Jake's. The world wants proof. Seeing is believing. And Jake wasn't here. I sure hoped it would happen quickly. Then I realized that I just got shot. It *will* happen quickly. I was lucky, unlike Yeshua. I shouldn't be afraid, but I was.

Although nobody booed, the Christian leaders were quick to challenge me. It was a perfect moment to show the world religion's disease. The man who stepped forward was none other than Billy Graham. He was the president's preacher. Hell, he was the country's preacher. He had more primetime coverage than anyone. I'm sure he was wealthy; TV wasn't cheap.

"Can you prove that you are Michael, the Archangel?"

"Can you prove that I'm not?"

"Well if you are, then show us a sign."

I held up my two fingers, the peace sign. The crowd broke out in a thunderous applause.

Mr. Graham was clearly frustrated. He pressed on. "Show us your power if you have the Holy Spirit. Turn this water into wine or raise the dead."

"You are asking me to do what Yeshua did, and not what is prophesied. Do you not know the prophecy of Michael the Archangel or the Two Witnesses?"

He appeared to be unsure himself. Of course he knew the prophecy. He preached signs of the end of times like all the rest. But he wasn't stupid. He knew I couldn't do any magic. He just didn't know what else to say. I asked him again.

"Of course I do," he replied quickly, and in harsh tone. It was obvious that he was getting angry.

"What am I to do then, to prove this to you?"

He remained silent. I felt sorry for him. He was like all the rest. He was conditioned to believe only if I could perform magic. And sadly enough, the Bible had a scripture for this defense as well: "There will be many that come as an angel in white, but let no man deceive you with all sorts of magic."

Believe no one! It was the same as signs. One scripture condemns those who seeketh a sign, saying there will only be one sign, the sign of Jonas. And yet they all preach the end is near and signs to prove it. The irony is that people buy it. Even though we've had wars, floods, earthquakes, famine, and droughts since the beginning of our history. And all in the name of religion. The beauty of Michael's prophecy is that it would happen to save the earth, not destroy it. It would be from a calamity that we hadn't known before. And this calamity was to be used by the kings of the earth when knowledge was worldwide. This was now. The world is threatened by nuclear war and we had the sign. The Jewish star represented that threat. It was the atom! And it is now in everyone's hands. My job, according to the prophecy, was to save it. I have to convince the world of my answer about us. The Two Witnesses solve the mystery of life and it tortures the world. I was torturing Mr. Graham. No, he was torturing himself. I didn't want to torture anybody.

Mr. Graham remained silent. The reporters asked him to reply. He couldn't. And I knew why. I didn't enjoy this. I answered for him. "You wicked and adulteress generation seeketh a sign, but I tell you there will be only one sign and that is the sign of Jonas."

The crowd broke into an applause. It wasn't what I wanted. I didn't do this to humiliate him, but only to prove how religion does condition us to expect magic. Their God is a blinking, thinking genie man. And yet the story doesn't support that. It's a story of scientific creation. "Please!" I cried out. "Don't rejoice at this man's mistake. Don't you love him? Be sad, for the day has come to see what religion has done to us all. I, too, am brainwashed. I still slip and say Jehovah." The crowd fell silent.

"Is there a Jehovah's Witness among you?"

One man stood up, no one else. "Before I start, I will ask you again, Mr. Graham. What am I to do?"

He replied weakly, "Speak the truth."

"Thank you for your honesty. And now I will ask you, Mr. Graham, before I ask the Jehovah Witness. What is the true name of Jesus?"

He stood silent. I asked again.

He looked up to the sky and said "I swear upon the name of Jesus as my Lord and Savior, to be the one and only name by which you can be saved."

I then turned to the Jehovah Witness. "What is your answer?"

"Mine is the same."

I then addressed the audience. "Is there a Jew among you?"

A rabbi stood up.

"Since you are of the same religion, would you please tell the world his name?"

He responded solemnly, "It is 'Yeshua'."

There was an immediate reaction from the crowd. They were demanding a chance to respond. The Jehovah's Witness held up the *Living in Paradise on Earth Forever* book. It was what I needed. They lost a lawsuit in my book over this very thing. I couldn't believe it.

"We acknowledge his Hebrew name, it

is right here." He had the book turned, to the very piece of evidence that would prove the name change.

Only this time it would be proven to the world on live television. It was perfect. "Can you please bring it up here. We have an opaque projector." I had seen it sitting on the table as I came in. What a coincidence.

The man came forward and offered me the book. I asked him to please put it on the projector. He did, and even focused the picture on the wall for everyone to see. "See," he said jubilantly. "There it is." He pointed to the name "Yeshua".

"Now will you ask the Jewish rabbi what the name is?"

He did. The rabbi again said "Yeshua."

The man became uneasy and pointed it out again. He even read it. "Even though 'Yeshua' doesn't sound like 'Jesus', they are the same," he continued on, "And the rest is history. And just like 'Yeshua', the name 'Yahweh' is the same as 'Jehovah.' And it is proper to pronounce the name. The name is the most important thing of all."

This was perfect. "Is this true, Rabbi? You are a Jew. Are there any Jews named 'Jesus'?"

"No," he replied. "They would be called 'Jeshua' in English."

The witness just stood there in shock. He wasn't ready for this. He obviously hadn't read my book or any other, just theirs. It was painfully obvious. He struggled to defend his position. He said he knew "Jeshua" was also correct. I asked him to point that out, then. He couldn't. They didn't write that part. I then asked him if it was all right to call them "Yahweh Witnesses." He said it wasn't.

" 'Jehovah' is the most important name of all," he said boldly as he pointed to that very sentence.

I stepped over and pointed to the sentence before it. I read it aloud.

"It doesn't matter whether you say 'Yahweh' or 'Jehovah'. They are the same," he said defiantly.

"Why not call yourselves 'Yahweh's Witnesses'? It is the accurate translation. We do have 'y's."

He stuttered and stammered.

"What is the most important name to you,

'Jesus' or 'Jehovah'?" I asked quickly. I wanted to finish this. It was *deja vu* all over again.

"'Jehovah'," he said loudly. He was wild-eyed by this time. He started to rant about my being a false prophet. I thanked him and asked to please let me finish this press conference. He wouldn't leave. Security escorted him off the stage. He was yelling now. He screamed all the way out that I was the devil. He made my point for me.

I turned to the Jew and asked him again, "What is the correct translation for 'Yeshua' in the Bible?"

He responded, "'Jeshua'."

I asked him if he was a Messianic Jew. He was. I asked him one more thing, "Is 'Jesus' the true name or false?"

"It is false," he answered.

The crowd grew unruly, and it kept getting worse. Soon they were throwing things. That was it. They rushed me off the stage. Mom was panicking. As soon as we got into the van she pleaded for me not to do this. She was scared and I was, too. But I had to do this. I wanted to tell them that the scripture "Heaven is God's throne and the earth is his footstool" is about us.

"We are them from the future!" I cried. "We've already conquered space and that is where we live. We know that's where our destiny lies. It's the safest place. Planets are prisons, and uncontrollable. In fact, they're easy to destroy. We know that we have to get off this one, if we want to survive as a species. We're rapidly destroying it. And now, with the nuclear threat, we can do it quickly. We can make it uninhabitable as a result of a global nuclear war. And the nuclear threat is global. We have to think of the future. Religion keeps us ignorant and divided. Alien evidence and time is what it's all about. Time travel. We've done it. What am I saying—we're doing it now."

I was crying and didn't even know it. Mom and Nurse Gail knew. They were trying to calm me down.

"Please don't cry, son," my mother pleaded. "It's all going to be all right. It's going to be all right."

Nurse Gail looked at me, and I knew it would. Who is this woman, I wondered. She was so calming to me. I did calm down and just held my mother as we pulled away. I wanted to know more about Gail. But right now, it didn't matter. My world was about to take off. I needed her. She could part Red Seas, I laughed to myself as I started to doze off. And believe me, we were about to have plenty of Red Seas to part. Religious people turned red with anger at my alien evidence. Especially when I said that making them bad only benefited them. They took the same stance on them as the devil. Even though they hadn't seen him, either. I was about to find plenty of evidence. Hard core evidence, like Easter Island. I found it in a book from *Reader's Digest* titled *The World's Last Mysteries*. Maybe so. But not for long! They supported my theory of where the gods/Aliens/angels live. I had an alien answer. They were giant heads with big eyes. They had on what the author called headdresses. They looked like flying saucers. I drew one beside it. They matched! So did the ancient drawings of flying saucers. Yes, ancient drawings of flying saucers in caves. Primitive man was telling us his gods/God lived in space on these space ships. The religious world wasn't ready for this. They had a spirit heaven that was pure magic. It was without science. But most of all, they weren't ready for what I would discover in November 2002. In *NATIONAL GEOGRAPHIC*!

But first, the book signings. Hell has begun! My book tour was scheduled. My music career and other projects immediately took a backseat. I continued to give all my royalties away. It was making an impact. Others were starting to follow my example. And in fact, many had already started before me. I became an advocate for futuristic thinkers. I invested in a worldwide recycling program. I started a think-tank on environmentally friendly products.

I enjoyed the feedback from everyone. I found that these people knew the importance of peace. It was the only way to survive. I became even more dangerous on my nuclear weapons stance. I voiced my opinion of total abolishment, starting with the United States. We had to set the example. This issue would become the most important.

Our situation with Iraq was coming to a showdown. We wanted them to disarm. We had to take religion out of our government. Our president couldn't see that his God blessing this country or having a chosen race was illogical and a violation of our civil rights. But he saw it with Iraq. Why couldn't he see this? I'm a scientist. My rights and freedom are being violated by his religious position. My son's history book pointed out that Moses was a killer, and yet they don't see the cruelty of this. We honor peace and know that violence begets violence, but we seek justice. We would soon encounter a Hiroshima on our own soil, but we forgot about Hiroshima. The cry for justice would take center stage, led by a Christian president. A president who follows a man of peace, but won't forgive, himself. I wonder if this man tells his children to fight back. Even worse, to make war plans! Can't we see the ridiculous contradiction here? The Islam religion was started by a killer. It was easy to see how they could be brainwashed to kill in the name of religion. The Jews as well! But why couldn't Christianity see the contradiction in their teaching of justice, pride, and favoritism. I was so frustrated at these book signings. I could hardly get any Christians to consider these points. And I certainly couldn't get any to see a contradiction.

Most of all, I couldn't get over the patriotism issue. It blinded people about the future. They couldn't see it as a stumbling block for uniting the world. Hell, we are part of United Nations! All of this was what I would soon lecture at the book signings.

I began my journey. My first encounter was in New York City. I missed *The Antonio Show*, and the world saw my assassination attempt. I regretted that, and had a hard time explaining why I rescheduled it to Little Jake. He and Angel didn't want me to do these book signings. We all cried a lot. But I couldn't turn back. Alien contact would eliminate all this religious war. And I was convinced of their existence. I had evidence. I seemed to be the only one who thought that we are them. I knew why they didn't show and I believed in time travel. But I had to convince my children of this, first. I showed Little Jake and Jimmy everything. They really were shocked at the tenth planet evidence. Even Little Jake heard about it in science and on the news.

"Why doesn't the world teach us about these clay tablets, Dad?" He asked.

"We're ruled by a religious government

that doesn't accept anything contradicting their teachings, son."

"That's not right, Dad. We're not religious. Don't we have rights?" Boy, he sure was getting an education fast. He sounded just like me.

"Well son, unfortunately might still makes right. And we are the mightiest. But don't worry, things are getting better. We've only recently become a global world, with the internet. It won't be long before knowledge will rule."

"I can't believe we haven't always had it," he remarked with great surprise.

"Me either, son. But I really can't believe how short our history is. We only have a recorded history of six thousand years. And unfortunately it's the Jewish history, which is what rules our world. Heck, That's only sixty generations ago, if we lived to be a hundred."

"That's an insult to my intelligence."

"Yeah, really! It's an insult to mine, too," I laughed. He talked so grownup-like. He was a treasure. I didn't get this opportunity with Jimmy. I wasn't a seeker of knowledge then, just money. Things were so different now.

I showed them the book with the tenth planet pictures. It had a lot of other pictures that proved our existence before recorded history. It had a picture of a satellite on a clay tablet. It depicted space travel between Mars and Earth. The man on Mars had on a helmet that looks just like the helmet of the face on Mars. Yes, we had pictures from NASA of a face-like structure on Mars. It had the same helmet. I showed them another picture of a space capsule in an underground silo. I showed them the medical symbol, yin and yang, Jewish star, and their scientific matches that can only be seen with an electron microscope. I told them about Travis Walton, and went and got his book. I read how Travis attacked them because of their ugliness. I told them my theory about us and our addiction to beauty. I told them how we hate ugliness because it scares us. Little Jake immediately agreed. I told them about Roswell. And finally, I asked them if they wanted to see the Alien Autopsy. They did. Angel didn't want Little Jake to see it. He was too little to watch it when I first saw it in '99. But this is different, now. He just saw me get shot. He needs to have rock-solid proof that they do exist.

She understood. I rented the tape. We all sat and watched it. I held Little Jake the whole time. He wasn't that scared. It wasn't done in a scary way. I was glad. They asked me how I knew it wasn't fake.

"We have ancient drawings of them." I showed him the Owl Man. I showed him Easter Island. I showed him the statues in the tenth planet book. The alien statues found all along the banks of the Jordan River. I explained how ancient man even named a bird the wisest because of its similarity to the aliens. I asked him which one that would be, and he said the owl. I showed him the Owl Man again. I explained why they named a bird wise instead of an animal.

"Because they can fly?" He was a smart little fellar. I told him they hadn't proven this tape to be fake and it is supported by ancient drawings and knowledge. Then I got a current issue of *UFO Magazine*. It proved my point about the Alien Autopsy. They were looking for anybody with proof of its authenticity. They also had an article depicting ancient historical evidence of flying saucers. I was shocked, myself. We had a painting on the Sistine Chapel by Michealangelo showing a man looking up at a flying saucer. It was a painting of Mary and Baby Jesus! More importantly, they showed cave drawings of flying saucers. I showed these to the kids immediately. I compared the cave drawing saucer to the one on the Easter Island head, and they were the same! I then showed them the Owl Man. I told them what I thought their messages were and how I could prove they were the same. "Also," I boldly said. "I will convince you that this confirms my answer about us and them."

"How?" They quickly asked.

"Well, let me show you." I turned to both of them. "The Owl Man and Easter Island. Do you remember my answer about them and us?"

"You think we're them," Little Jake answered quickly. He was highly competitive and loved *Jeopardy*.

"Yeah, but it's a little bigger than that. I think we're them from the future. And we re-addicted to the beauty of mankind because it gives us power over one another. The more beautiful you are, the more power you have."

"I know that," Little Jake replied. "Just

come to my class and hang around. You'll see for yourself."

Man, I thought to myself. Truth sure does come out of the mouth of babes. I laughed and said, "You're right son, children can be cruel. But they don't do it intentionally. Well, some do." I quickly remembered little Shirley. She was the poor little girl that I always took up for in the third grade. I was new then, and she was my first true friend. We sat beside each other and lived real close. I even walked her home sometimes. But it was hard. The kids would tease me about it. I got on to them later. But I sure knew how some kids could be cruel. "Look at these both," I said to everyone, continuing on. I knew how short the attention span of kids could be. "Do you see any connection to their messages?"

"Just show us," Angel replied quickly. She didn't like trivia. Good thing the kids did.

"Well, do you see any connection to where the Owl Man is pointing and where the headdress on the Easter Island head is?"

"Yeah," Jimmy replied first. He beat Little Jake to the punch. "They're are both up."

I laughed and said, "See, our competitive

nature is proof of my answer, as well. We love to win."

"All right," Angel snapped. "I've got things to do. Get to your point."

"Well, they both tell us that their gods live in space, but the Easter Island statue shows us what in. Spaceships! And check this out," I said with an excitement that was hard to contain. I pointed to the Cernes Giant. "This is the statue that answers why they stay away from us. We are addicted to the power of man's sexuality." I pointed out that the giant was holding a club, symbolizing power, and had an erection symbolizing our sexual condition. "Beauty of the flesh rules! And can you only see these statues from the sky, boys?"

"How does that prove that we are them?" Angel asked immediately.

"Well, we've only been able to fly for about 90 years. We've only been able to see these symbols with an electron microscope for about 70 years. We have Roswell! And we've just now seen an alien body. I'd say that pretty much answers your question, doesn't it?"

"I don t understand," she replied.

I explained that these images of science and what we now see from the ground weren't possible by a primitive culture without these tools. She started to get it. Then I quickly pointed to the Owl Man.

"Look, if art imitates life, and it does, I'm an artist, then this is a statue of what their gods look like. It answers where they are and who we are. One arm up for them and the other down for us. We're the only ones here. So we must be them. And all these scientific symbols matching their religious symbols proves the possibility of a scientific God that has created us scientifically and gotten addicted to us. Heck, do you want to be any uglier than what you are?"

"No," she snapped again.

"Neither do I. And I would bet nobody else does, either." Both of the boys immediately agreed. "You see," I continued. "We all know it. And the Cernes Giant is the best proof of all. Hell, I even think that the aliens made it. Look how anatomically correct it is." The boys laughed.

Angel didn't. "Well, how does this prove we exist forever? How does it prove your time travel theory? And even if it does, I don't want you to do these book signings. You're going to get killed!"

"Honey, please help me do this. I need for you to be strong and reassure the kids that everything is going to be all right. You can read about my studies of Albert Einstein in my book. We're made up of atoms; you know what the first man is named in the Bible, and what Einstein split to give us nuclear science. And most importantly, what the Jewish star represents.

"I've got to do this. I believe in Einstein, the theory of relativity, and time travel. I believe in Yeshua. And I believe I could be Michael, stuck in time travel. I believe anything's possible. I would die for my sons' futures. If I can make it better I would. I've got to, baby. Please help me? If they kill me, I'll be back. I promise."

"Nobody else has come back," she replied. "And a lot of people said they would. What about that crazy couple in California? They didn't come back."

"No they didn't." I remembered Ti and Do. They called themselves the Two Witnesses, just like my book. "But, what the hell does that mean? Almost every Christian religion claimed to be them, even the Jehovah's Witnesses. But none of them were saying the Jewish star is the sign for contact, to save the earth from nuclear destruction. Most importantly, they don't acknowledge that mankind is evil. We are sexual! We are different-looking."

"This can't work. Come on!"

"Please honey, you've got to believe the evidence. It's the only thing on this planet that doesn't lie. I'm just a scientist who believes it. I will keep looking and I will find more, but you've got to believe me. Please?" I hugged her. We all hugged. They all knew this was my destiny. Alien evidence! I was going to find more.

I went to my first book signing. It was a live broadcast by *The Antonio Show*. It was held at the biggest Barnes and Noble on Times Square. It was spectacular. I've never seen so many people. Not even at my concerts. This was like New Year's Eve. I did the book signing inside, at the persistence of my family. My bodyguards were present, Mom and Nurse Gail. They did have it playing on the monitor at Times Square.

I answered all the usual questions about where heaven and hell were. Then, I had to explain that the atomic symbol really wasn't an accurate model of the atom. But I quickly explained that it was the universal internet symbol. It's because we were taught that an atom consisted of a nucleus surrounded by an electron, neutron, and proton. I brought my son's fifth grade science book to prove this point. It stated that very thing.

I had one Christian scientist say that the symbols and statue references that I made didn't look anything like what I said. He stated that anybody could make theories, but it didn't make it so.

"Yeah, I agree. That's my point with your Bible contradictions. But do you agree?"

"Absolutely not," he declared defiantly. "It is the infallible word of God. Would you agree that it is written by men?"

"Yes, but God inspired it."

"Why didn't God just write it, himself? Then we wouldn't have all this division among you religious people. Or better yet, why doesn't he just show up? Is it too much to ask a loving father?"

He immediately got angry and told me that I was going straight to hell. I loved it. This guy was my best salesman. "Oh, by the way. Would you do me a favor and draw DNA beside the medical symbol, and then a UFO by the headdress on the Easter Island head?"

He couldn't back out. He needed to repair his image. After all, he was a scientist who taught physical science at a local Christian college. He drew the objects and they were just like mine. The audience responded with a resounding acknowledgment that they were the same. He hollered that they weren't. I asked him if he ever wondered why this mystery even existed anyway. He said life was a lesson.

I said it was a mystery and we just haven't figured it out yet. That's my point of giving my book the title *The Two Witnesses and the Religion Cover-up*. It was the point when the mystery was solved and it tortured people. I think it's because we were ugly and powerless as aliens, but I could be wrong. He said I was. I asked him if thought he was right. He did. I asked him if he thought that people could be religiously brainwashed. He did. I asked him if he thought he was or could be. He didn't.

"Well, you might think again. You just exhibited the first, second, and third symptoms of religious brainwashing. You think you're right, and you're religious. You think others can be religiously brainwashed, and last but not least, you're religious, but you couldn't have been religiously brainwashed. Finally, I give you one last thing to ponder. Just some food for thought. Did you ever wonder why your God takes time to do things when you teach a thinking, blinking God? And most importantly, if his teachings are about the inside and he stays away from us, isn't it possible the outside appearance of him is ugly? And if that doesn't give you some doubt about your righteousness, then aren't you sad for us? I mean, you seem to be mad."

He didn't even realize that he was standing over me with his fist clenched, and was red as a beet. "I'm not mad, and I do feel sorry for you."

"Well, if you really do, then please unclench your fist. The whole world is watching you." I pointed to the television.

He relaxed and smiled weakly. "I'm sorry," he said in a low voice.

"That's all right," I replied back. "I feel sorry for all of us. We've all been brainwashed to have a God that isn't here, can make the world perfect if he wants, and even sacrifices his own son. That's a shame. And we call him a loving God. We are shameful for not challenging these things. Things like racism, favoritism, killing, abandonment, incest, and downright torture. It's all in there, and we ignore it because religion rules the world. If you don't believe in God and country, you're ostracized. That's the sad reality of our condition. Besides, aren't we all his children? That's something you should be able to answer. The Immaculate Conception sounds like alien abduction to me. But that's my opinion. I could be wrong. It's just one of my many scientific discoveries that I've made. Like the alien statues, the scientific symbols, Roswell, UFOs, and Easter Island. And most importantly, God killing his children when he doesn't have to. That's enough for me to never think that I'm right. It's just about discovery. And no offense, but I think religion ought to be banned. It's a disease that keeps people from being open-minded and neutral. I can only hope it doesn't affect your teaching habits as it has our president. It is a violation of our civil rights. That's an indisputable fact."

I held out my hand and he reluctantly shook it. He calmly walked away. But he couldn't control his religious disease. He said he hopes for my sake that I find the truth. His truth.

I wrapped up the signing. The crowd was in a frenzy. It was mostly boos. There were a few things thrown. I hurried into the van and we drove away. I was determined to study Einstein, and to keep honing in on proof of time travel. It was my main theme. There was plenty of evidence that it had been conquered already. We're the time traveling machine.

I couldn't wait for the next preacher. I was determined to quote "Many are first that shall be last, and last that shall be first," from then on. I felt sorry for us all. We're obviously all addicted to man and power through beauty of the flesh. I sure loved my discovery of reincarnation and how it's taught by all cultures on earth. I loved the definition of the atom, the immortality of it. I loved its creativity. But most of all, I loved its infinity. "I loved its infinity," I said aloud. Then I thought of its symbol and how it looked like a cell dividing.

"What did you say, Mike?" Mom suddenly asked me. It got my attention. "What did you say? I asked you about dinner. Angel wants to know about dinner. She wants to celebrate."

"Celebrate what?" I asked in a daze. I was still thinking about all the things I wanted to ask the Christian preacher who said that he's a scientist. I wanted to tell him how I loved the atom. Anyway, I thought to myself as we sped along, trying to lose the followers, or the ones who hated me. I didn't know which. I knew Angel really wasn't joking, though.

"You're not getting killed," she replied quickly. "Mike, you do say some things that are very dangerous." I gave her that surprised look, just to lighten up the seriousness of the truth. I was getting death treats. "Michael, you know you do, now. You need to cool out a little," she smiled and poked me in my side.

"I know I do, Mom. But I'm just pointing out facts. And I love the fact that reincarnation makes sense. And it is supported by the atom's definition. We are made of atoms. They're immortal and create with time. Life, death and rebirth. We must be born again of our own immortality. Spirit, the atom, life. The word is energy. Something invisible that creates everything we see and don't see. We can never stop existing, just like the atoms we're made of. Macro science says that the smallest part of the universe is exactly like the largest part. It's all-infinite. That's what I was quoting earlier. I loved its infinity, the atom. And the first man is Adam. Maybe we're infinite." She looked surprised and a bit confused. "I am just pointing out the contradiction to your religious teaching of one chance for man. Then it's a reward or nothing. According to them, you just stop existing. I love this quote from Yeshua that I use as an example to support his belief in reincarnation and science, 'Many are first that shall be last and many are last that shall be first.'"

"That's still hard for me to understand," Mom said bewildered-like.

"Well don't worry, Mom. You understand another quote that means the same thing, 'Reap what we sow.' You live by it. I watch you. And I love you for it. You can forgive and love even those who hate you. And even those who beat you and your children."

We looked deep into each other's eyes. She knew my pain and I knew hers.

"That's karma, reincarnation and Newtonian science. You know—Sir Isaac Newton. 'For every action there is an opposite and equal reaction.'" She looked somewhat unsure of the name, but she knew the law. "No offense Mom, but it's only because you are uneducated. And actually the word for that is ignorance. I don't mean it in an insulting way."

"No, I know you don't. I am ignorant," she agreed, quite happily-like. She was so easy going. A real happy-go-lucky person. I loved her so much. She was my light in a seemingly infinite sea of darkness. I laughed to myself. She caught it. "What?" She asked with a suspicious smile.

"Nothing, I just can't say that to anybody. You're so easy going. That's such a rare and beautiful thing. In this infinite sea of darkness called space. Heaven," I said and looked at her smiling with admiration.

"Well, I'm just being honest," she remarked again. She didn't want to appear self-righteous. She knew that I was more educated than she was and did study more. She didn't. She had only studied their material. And she just had an eleventh grade education. She knew that I was just a fact finder. She really loved that about me. She enjoyed my college pursuits. My dad did, too. I think we all have dreams. But none of us like know-it-alls. Yet we all like knowledge. It gives us power. Everything is about power. And family love is the most powerful thing of all. I knew that with Mom and every time I looked at my children.

"Mom," I laughed. "You couldn't be dishonest if you wanted to be. And besides, when you say that you're just being honest, you're implying that you're not honest all the time." She smiled real big. "That's funny, huh? See how easy it is to say something illogical?"

"Yeah," she said laughing. "We all do it, don't we? I never thought about it, though."

"Me either really, until lately. Now, I have to. If I don't want to look ignorant," I said laughing with her. Man, I really love Mom.

Nurse Gail just sat and laughed the whole way through our conversation. She finally broke our chain of thought and snapped me back to reality. "Okay—enough with your observations. You better start thinking about dinner and Angel. Or you'll have another force to reckon with," she giggled. "It just might be the dark force, though. You know."

"Yeah, hell," I replied quickly. Mom just laughed and got onto me for saying it. But she did laugh. To Angel, dinner was everything. I knew that. After all, she used to be a model. Just because I was a guru, didn't make her one. I was the asshole. But I was trying to change things. I gave up vegetarianism. And wow, she was right about my not getting killed. The crowds were getting bigger and angrier. This was just the beginning. My first signing, but not my last in New York City. I would do one more.

The next year flew by. I would get threatened at every one. I continued to get death threats in the mail. But I never got physically hurt again. Not yet anyway. That was to come in 2012. In NEW YORK CITY! I knew it, when the Olympic committee announced their winning bid later that year. It was the beginning of the end. The end of our time. We would soon threaten it with nuclear war. New York 2012? That's what I felt.

I sure hoped all this was true. Contact, that is. I knew war was a certainty. The whole world did. I had faith, if nothing else, in uncertainty itself. The mystery wasn't solved yet. And with today's scientific advances, I felt that anything is possible. And contact seemed possible. The definition of the atom made it seem so. It was an endless world full of endless possibilities—even preventing the end of us, which seemed so inevitable. That was what faith was: belief in endless possibilities. The mystery of life itself. Mystery!

Do nothing and life will thrive. But we were ready to destroy it and ourselves.

I had faith in my alien discoveries. Alien evidence was everywhere. The evidence suggested the future has already happened. It appeared to be unfolding again before our very eyes. And perhaps again. I always remarked on the evidence of *déjà vu* and time travel. I often wondered why people didn't see it. Are we really the disease of the universe, destined to be kept on Earth until we threaten our own existence? Then, they save us from ourselves. I always felt it. I started seeing it. Especially when I saw the Easter Island heads in the book called *The World's Last Mysteries*, by *Reader's Digest*. They had alien eyes and headdresses that look just like saucers. And big heads! I was showing this to everyone. I got a recent *UFO Magazine* showing the Alien Autopsy. It reported on its authenticity and put out a reward for anybody to disprove it. They didn't. I got another one on aliens, UFOs, crop circles and other ancient evidence in our past. It blew people away. Especially the cave drawings of UFOs that look just like the saucers on Easter Island heads. There was a cave drawing of a man in a spacesuit that looked just like ours today. There are ancient paintings of saucers in the sky. One by Michelangelo, of Mary and Baby Jesus. There is a man in the background, looking up at a saucer. It looks like the Easter Island headdress. It was shocking to me. They even had the Moth Man. It was yet unsolved. And the shroud was still unsolved, just like spontaneous combustion. Then, I found out about the star child skull. The alleged alien skull. It was from Peru. The same place as the Owl Man!

I was showing all of these at the book signings along with my alleged matching symbols. It sold books and made many Judeo-Christians angry. The Christians were really starting to rally the demonic/alien/spirit movement. It was growing. And then it happened.

November 2002! *National Geographic* showed an alien-looking mother goddess statue, found in Israel! I started writing the next book. The first book lacked this evidence. This was the seeing-is-believing evidence. They were flesh and blood, and the Christians couldn't deny this. This evidence was supporting a scientific creation with my matching symbols! An alien statue on a mother goddess body. It was proof of what the gods looked like and that these religious symbols were scientific. It would put the nail in the coffin lid on the invisible spirit God who was a thinking, blinking genie. Now, the yin and yang won't be good and evil anymore. It's two sperms in an egg, and we are the scientifically created missing link. We are the unusually large-headed mystery. We are really evil, though, I laughed. To myself, it was simple. But really it's all so twisted.

With all of this I was prepared to make my case. I felt that it would convince the world. Well, except for the religious world. It would make them angrier. It was all about now! Science was winning the struggle. Physical evidence dictated theories.

What's present and before us is gigantic. Alien evidence and religious nuclear war. We were about to strike Iraq. Korea was threatening to strike us. We were struck. 9-11! It was terrible and yet not nuclear! That would come. I must stop it, but I can't. How do I protect my sons from it? I can't perform any magic. I'm scared. I'm so scared.

I started putting this book together. I knew seeing was believing. I put the alien statues on the cover of the book, along with the atom and Jewish star. Then the tenth planet and the ancient depiction of our solar system. Also, the pyramid and the year 2012! I put as much as I could. I needed to help ease my children's fear. Even kids knew we couldn't stop the inevitable. Nuclear war! The most important thing to Little Jake was my name. He knew I was Michael. He might not have understood the prophecy, but he knew about nuclear war and its threat to end our lives. Hell, everyone did.

But I was the scientist who matched the atom to the Jewish star. It was on the book cover. There was a problem, though. I didn't know if I was Michael or not. I mean, I haven't been told. There wasn't any contact telling me so. But I did belong to D-Company Platoon 2012 in the Marines. That was weird. Will it be D-Day? It was all too weird! But in a bittersweet way, an uncertain way. You know, the way you love your children and want them more than anything. But you can't explain why they have to die.

And why would you do that to them, anyway? Why'd we bring them here? They're scared! We are too, but we want them! And I even let Angel have another, a little girl. We named her Hope. She was about three now. Old enough to ask why! It's now or never. The present. Their present will be my sacrifice.

My mind was racing with thoughts of this next book. A large part of it is the future. Do we have any hope? I knew we did, but not with mankind. Only with the aliens. I remembered the scripture

"Seek first the kingdom of heaven and store not your treasures upon earth, but in heaven."

Our hope is with them. I will sacrifice myself. I must for my children's sake. For the sake of all children. I am afraid. But I love them. This is written.

The future is but a dream. And dreams can come true. I thought about the dream. The Two Witnesses. It was our time! It wrote itself as a result of my seeking and discovering. I continued to discover. I hoped I would never stop. It all came true. Maybe this book would, too.

Now to conquer time, itself. That was a mystery yet to solve. And if we could, it would be a dream come true. I was a dreamer. "Row, row, row your boat gently down the stream, merrily, merrily, merrily, merrily, life is but a dream." I remembered how Mom used to sing that to me as a child.

I learned on the guitar "Michael Row the Boat Ashore." But I'm afraid these waters weren't going to be gentle. And I wouldn't be rowing, I would be running. Life wasn't a dream. It was real. It was scary. It was a lot like my dream. It was a nightmare that would lead me to my inevitable death. And now for my future rebirth. To quote Vince Gill, "There's no future in the past." Now, I will ask for his help and everyone else that is wealthy. Give it away. I will. Will you?

The same answer applies to our nuclear dilemma. We must give them up first. You can't enforce peace with violence. And we can't make all men good. We must believe for our children's sake. For a better world for them. Peace is the answer. The one final thing that I would put in my book was the Egyptian obelisk. It looks just like a nuclear missile and is even called a rocket ship. The author of the twelfth planet supports this, which is about our knowledge of a tenth planet in our solar system.

THE PRESENT
DECEMBER 1, 2002

It was a cold winter day. I was signing in Nashville. I hadn't been there long, before heading to the bathroom. I drank too much coffee. I stopped by the magazine rack to check out *UFO*. I got the *National Geographic* instead. I had to do the number three. That was one and two combined. I laughed to myself. I needed to humor myself. I was about to be criticized, insulted and attacked.

I was blown away when I started reading and saw it. The author didn't see the alien resemblance. Even though, she described the long, slanted eyes and bulbous head. I couldn't believe it. I wrapped up a quick signing and headed home. I had to finish my next book. I also had to contact the magazine. I tried to contact her. I used this picture. They never contacted me. I really didn't care. If they sued me I would gladly give them what they want. I would do anything for the sake of scientific discovery. Except to kill my children. I left that up to religion.

Was she religious? Why couldn't she see the resemblance to the aliens? I read the article again. The archaeologist who found it was Yosef Garfinkel of the Hebrew University. Huh, Joseph. I bet he's religious. What a tragedy. It seemed like the whole world was religious. Even the *National Geographic*. I knew how media sources could be. The Nashville newspaper wouldn't give this a chance. It would come through a small newspaper in Cookeville. And man, would it get things cooking.

I had to get the book done. We were about to go to war with Iraq. The year ended with one religious suicide bombing after another. It began with Michael Jackson and the Columbia disaster. It was the challenger all over again. *Déjà vu*. It was horrible and exemplified the reality of mankind's existence.

The space agency needed money, and Michael Jackson had it. A lot of people had it. Our government. Yet, we put more money in preserving ourselves independently than coming together. And we needed to come together to conquer space. The news of the Columbia's disaster took a back seat to Michael's challenged spirituality.

What a sad joke. Like we need to challenge any wealthy individuals' spirituality. He said that if it weren't for kids he would commit suicide. I wondered why he didn't give his money to NASA for our kids. We do need to get off here for their future. We need to conquer space. Michael was spending lavish amounts of money on gifts for himself. He was just like Saddam Hussein. He had paintings of himself hanging all around his palatial home. He was spiritual, though.

Right!

I often wondered why I used his name. Was it just because of our similarities, as I suggested earlier? Or was it because it is written? This was easy for people to see through. I would like to challenge him, first. I want him and all wealthy people to give their money away. I was dangerous! I was ready to ask the world to stop and think about a God who chooses favorites. A God who controls all of this and yet, doesn't stop it. Or one that would create it anyway! A father who blesses one child and tortures the other to the point of death.

I was about to challenge our president. I wanted him to stop saying that we are blessed and asking God to help us in Iraq. He's violating my rights and my children's. Just because he doesn't see his illogical God story doesn't mean that I don't. His loving father created children knowing that he would kill them.

Come on! Do any of us think we can stop nuclear war? Nobody stopped us the first time. We bombed Hiroshima and Nagasaki. Our country killed hundreds of thousands. They were all innocent! They weren't soldiers. And even if they'd had the bomb, they wouldn't have bombed us back. Violence begets violence. Do these religious leaders really believe in life after death?

I can't believe this and I can't stop it. All that I could soon do is to ask questions of them. Can you forgive me?

We have seen and will again see the results of their religious brainwashing. I kept signing and again was chased away by religious zealots. They wanted to save me, though. I guess by killing me. And they said they love me. They were just like their God, loving and killing. Don't they see this?

I almost had the book finished and got my first signing. It wasn't at a bookstore. It was going to be held in an open parking lot. The bookstores were too religious to host it. I talked a pool spa and depot store into it. It would be perfect. I wanted everyone to have a good time. The party was on me. We needed the break. The world was about to go to war.

The newspaper gave me my article. March 15, 2003! The signing was scheduled a week later. I couldn't believe it. It took a lot of guts for the young lady to put her career on the line and print my story. I was now challenging the Israeli scientist who discovered the alien/mother goddess statue. He taught at the Hebrew University in Israel. I also challenged *National Geographic*, the president of the United States and the world leaders to consider this evidence. After all, alien evidence would eliminate religious war. It would eliminate division. It should bring us together and cure the world of the deadly diseases called patriotism and pride.

The article got the national bug. But it didn't get me any respect. People were quick to point out that I didn't have a degree in science. In fact, I was dropout. I didn't finish my fourth year of college. I failed the last semester.

My son Jake let me know it, too. I told him that I did have a two-year degree. It was an associate of applied science. And I did take logics! "Okay," he said. "But you still need to finish."

I knew I did, too. I wanted to become a quantum physicist. The atom was consuming my every thought. I was reading everything I could, from Albert Einstein to Stephen Hawking. His book was the number-one seller.

Everybody wants the answer to the mystery of man and the universe. Could I really have the answer? Well, at least the mystery of man, anyway. I believe you have to learn about the universe by seeking. And I didn't even compare to Einstein or Hawking. The evidence of gold was pointing toward the aliens and the reason the mystery of man existed in the first place—for power. We must be them, and this must've happened before. We needed gold for making more earths.

We can't cure others who have an addiction to power, only our own. And man will always be addicted to power. Power seems to be the reason they established a hell. The reason the aliens put life back on planets. We were condemned for our desire of power. The Eastern religions say power corrupts the soul. We tell children to share and turn the other cheek. We know the evil of power. And now we have a dilemma. We call ourselves the most powerful nation on Earth and we're proud of it. We even have a president who says we're blessed by God. Religion was holding back discoveries. Holding back progress.

Are we them? I had to know. I couldn't do anything magical if I did.

That was the tragedy of the future, which is about to unfold. We are always about to encounter the future. It has begun, it began, and it is beginning again. Stop, DO nothing, unlearn! "Enter into my rest." Peace is the only way. I held up my two fingers and thought about the Two Witnesses. Coincidence? It was a universal sign.

The FUTURE!
MARCH 2003

"The eyes of March are upon the world."

I am stopping myself and am taking a break with my wife. We are watching a movie. Life imitates art. We love movies. This one is called *Road to Perdition*. I laughed at first and then sat down to watch the movie.

It was just like life. Even down to having a Michael at the end, who would ultimately be the peacemaker. His father was shot and Michael pulled a gun on the shooter. However, he couldn't do it. His father shot the man as he died. Michael resisted revenge. He saved his father's dream, even though his father's end seemed to justify the means. But in the end, he knew it didn't. He didn't want Michael to fall prey to his same mistake. And Michael didn't. Michael would live the truth. It was just like the Bible story. Maybe my story.

I was on the verge of tears during the whole show. I felt the same pain and love of life as they did, in raising my sons. And I was shocked at its real-life depictions of hell that innocent kids and loving parents go through. But now I understand the saying "The road to hell is paved with good intentions." This movie captured that hell beautifully. Perdition means hell.

Hell, I understood it. I had beaten my son. My brother had committed suicide. My dad beat my mother. I wanted to right all my wrongs. I wanted to rid myself of shame. I lived hell. I created hell! My road to hell had been paved with good intentions.

But it was time to make sacrifices of myself. I could only do that by doing nothing because it was doing everything. I saw nothing good about mankind, and I was one of them. I did see our necessary evil purpose, though. We needed gold to contain ourselves. We were the perfect workers. Robots don't compare. But I must have hope. I was entranced by the power of the movie's ending, when I realized that I did have Hope. She was coming toward me with those arms held wipe open, wanting me to pick her up.

"She must've been awakened by the movie," Angel said as she reached for her.

"Yeah," I replied as I picked her up and went on out of the room. I needed to air my eyes out. I was a softie. Right then, I knew had to finish a song that I was working on. It was one of only a few that brought tears to my eyes. The opening line was "Why do I cry at the movies?"

Angel followed me. "Here let me have her, go write your ending. The signing is Saturday."

"Wow, is that a coincidence or what?" She looked puzzled. "You know, the cover-up and Saturday being the real Sabbath."

"No, it's just the day you always do signings. Duh!" She didn't get my connection. Or maybe she did. Did she know about the song, too? Anyway, she took Hope. Then she said, "Besides, I know how you cry at movies,"

Wow, did she know what I was just thinking? "Yeah," I agreed and went on to the computer room. I didn't want to engage in further conversation. I needed to finish this book. It was my road to perdition. I saw man and the nuclear dilemma as the ending. Our road to hell is nuclear. And I couldn't stop it. Therefore, I finished the last chapter. I used all the pictures. Without permission. And I said all these things on the road to perdition. I knew it was the right thing to do.

We were about to attack Iraq. I wasn't about to predict the next nine years. It didn't matter. It seemed all too clear that I wouldn't stop war. Nobody would. The aliens would stop it by resurrecting the Two Witnesses! It is to take place after the war involving an army of two hundred million. This was the same number as China's military forces. They are a nuclear power. But, who are the Two Witnesses?

I will sleep one more night before I write again. I do not feel worthy to be one. I have done wrong. I beg for your forgiveness. I beat my son. Tomorrow I will wake to the mystery and then finish this book. But, whatever the outcome, my observations are just that—observations. I do not belong to a religion. I am a seeker. I want no followers. I am getting ready for the book signing. It is my first with the *National Geographic* statue. It is my first as a lawyer for the alien case. These are the things that I must prove—that the evidence will prove they are the gods of primitive man, their motive behind our creation, and why they don't show.

I went to the signing. It was televised worldwide. I would not answer any questions. I started by saying that I don't claim to know everything and I certainly could be wrong. However, I am presenting a case for the alien existence. I propose to answer the mystery of man alone. I am not educated enough to answer the mysteries of the universe. I am studying quantum physics. This is where I believe those answers are.

I apologized to those who had bought my first book. "I was very angry with the Jehovah's Witnesses and religion in general. I am sorry if my story offended any of you. I think religion is a form of traditional brainwashing that should be banned. We don't make our kids join a club to be good. We make them accountable for themselves. We can do the same for ourselves. If your religion can't answer every question in life, then maybe they aren't 'right.'"

That's a safe assumption. I use my former religion as an example. "It is judgmental by nature, and self-righteous. And they can't answer every question. That is a fact that they will gladly accept as true. They think they're right and everyone else is wrong. However, I would like to start my case with proof of their ignorance and fraud."

I put my exhibit on the opaque projector. It was evidence from their own material. I first pointed out the fraud in Yeshua's name change. Then immediately finished with their answer to the $64,000 question. "Where is God and what is his purpose?"

They said he was timeless and always existed somewhere in universal space. I laughed and asked the audience to please tell me why I just laughed. A lady out of the audience immediately said I thought God was everywhere and invisible.

That was my point exactly. "Thank you," I replied.

I didn't mean to laugh because this is not funny. This is the crime of religion. They base this on nothing, not even a scripture to support it. And in fact, as the lady just mentioned, they teach individuality, not omnipresence.

This is the opposite of it. I will not elaborate any further on their God's purpose. Although he is proud at the end of his creation, he wins the game of torture that he foresaw, and could have eliminated if he wanted to. But he likes competition even at the expense of killing his children. I will say that this invisible, omnipresent God is the definition of the atom. I explained all this to my parents. They didn't really understand. Right!

I asked my mom if she thought she could be brainwashed. She didn't. I told her I did and was, by her and dad. I told her that I didn't blame her or anyone. This is the tragedy of religion.

Now, I would like to make my point in a humorous way. The same way I did for them because it made my dad angry. Anger, I told them, was a mask for ignorance. Ignorance and anger are products of religion. This is a fact. My dad proved it. The Jehovah's Witnesses proved it. The world proves it. Blind faith is the easy path.

Anyway, I used the Jeff Foxworthy approach.

"This is it. If you are religious and think others can be religiously brainwashed but you can't, then you could be religiously brainwashed. If you deny this possibility, then you are exhibiting the first symptom of any addiction—denial! And you are religiously brainwashed, and we all know the first step in the twelve-step program. Confronting our addiction through admission."

I always waited for a chuckle. If I didn't get it I usually ducked and covered. This is the reason I began to get educated. Religion is deadly. You don't have to belong to it to be good. My scientific observations of life led me to this answer.

"The reason the mystery exists is rooted in the addiction to power. Life is an addiction. Power is self-destructive. If you feel good, you are experiencing a sensation of power. From a drug, to food, to sex, it's all the same. It's about power!

"It is the reason for the fall of the Jewish angels to the Greek gods. I theorize that religion began when the gods lost control of their creation, us! This leads me to my first and second goal. Proving that they are the gods of primitive man, and answering why they created us. I will do this with the physical evidence that I have presented in this book.

"First, we will look at the front cover of my book. It clearly shows a mother goddess statue with an alien head. It is older than any Biblical writings by man. It was just recently dug out of the ground in Israel. A man who believes in an invisible God dug it up. He is a Jew. The statue provides evidence of what these gods look like.

"It is the alien head. Other statues and art support this theory from all over the earth. They are all older than our recorded history from the Jews. The evidence that proves their flesh and blood reality is also evident with the alien skull of northern Mexico. Their story is the same as the Jews'.

"Except for the spaceship element. That's how they came here. The Jews make it a spirit thing. However, the Jews make it clear why they were doing this. Because the daughters of men were pretty. They begat giants.

"This is in every culture. Hercules was half man and half god. Alexander the Great said he was the last in the lineage of gods and man.

Then we have the Cernes Giant statue itself that clearly shows the power element of this story and the sexual. We have the Blythe Giants in Southern California. One of them has a huge head.

"I have discovered the DNA symbol on every continent, along with the atom. I put a Mayan pyramid on the back cover to support this. We have the medical symbol from the Jews and the yin and yang from the East. The yin and yang supports this hypothesis. It clearly shows their sperm/DNA and primitive man's together in an egg or cell. The Hopi petroglyph shows a god with a big head controlling the five races of man. It is approximately 600 years old. How did this shaman know of the five races of man, let alone the alien existence in New Mexico?

"The petroglyph shows six beings. There are six steps to the Mayan pyramid. Man was created on the sixth day. I am the sixth son. The atom and the Jewish star are a six-pointed symbol. Coincidence?

"The Hopi shaman only saw the red man and white man. Or did he? How did ancient primitive peoples have these scientific symbols without us finding an electron microscope?

"The Mexico legend of the sky people also tells us where they lived. They lived in spaceships. This is supported by the headdresses of Easter Island. They look like today's flying saucers. But we also have ancient evidence of flying saucers in caves and religious paintings. The most famous is of Mary and Jesus. The saucer in the background is identical to the headdress. Even the Biblical scriptures support this fact: 'Heaven is God's throne and the earth is his footstool.' The saucer is on top of the Easter Island head for that reason. It indicates up. Space is heaven. The Owl Man is pointing up. It has a big head and big eyes. That's the reason scientists call it that. It is the god of the Nazca Indian. It looks just like the aborigine god, the wandjina of AUSTRALIA!

"They are on the opposite sides of the earth from each other. They look just like the Hopi's god and the head on the mother goddess statue. The statue on the banks of the Jordan River looks just like them. But it is the most definitive of all. They show both sexual organs, male and female. This sheds light on the Biblical scripture 'In the beginning we were created male and female.' It is on the back cover of my book, below the saucer.

"It is clear that the mother goddess statue not only shows what our gods look like, but also that they are flesh and blood. I theorize that we are the products of this creation. Modern man is the missing link. Our five races support my competition theory of power. This is my evidence for the aliens being the gods of primitive man.

"Before I go to the motive behind modern man's creation, I would like to clarify my findings. I only found one type of god. I did find evidence of other strange beings. I theorize that these reptilian beings on cave drawings and statues are indicative of their scientific pursuits in creating the perfect worker—one who can create itself and reproduce. We have many mysteries to support this and even historical evidence. The Sumerians give clay tablets showing these creations, as do the Greeks. We have the centaur and the Moth Man. The Moth Man can fly and has big eyes. However, I have concluded that the statues, drawing and cave art of the Roswell alien is the god who created all these other genetic mistakes. Even their singleness gave rise to the oneness of religion. They all looked the same, as is supported by the evidence. What created them is invisible male and female. The atom created the Adam. And both are really invisible. Science verifies a Jewish scripture 'All that is visible is made with invisible hands.' This should explain the science behind monotheism being the same as polytheism. It will hopefully make it clear about the history of the Jews. They began as all the others, polytheistic. Their story of man starts with heaven and the earth being one, void and without form.

"INVISIBLE! Man began as one, male and female. The fall began with two trees. One was forbidden. It was the knowledge of good and evil. I theorize it is us. And they created us with the knowledge of DNA, two strands of information. A Jacob's ladder of genetic information. It all begins and ends here. The WORD!

"Now I will address their motive for our creation and why the mystery continues. It will provide a logical scientific reason to the question of why the aliens don't show. All that we need to do is follow our physical evidence back to our beginnings.

"What is the driving force behind our existence? It is money! We even began to populate faster was we became more civilized. We became more civilized as a result of a monetary system. And we know that money is the root of all evil. Our civilizing is evil and destructive to the earth. This is supported by Yeshua's teaching about the cares of the world, and the Eastern religions with desire. The Hopi petroglyph even shows the ones that will inherit the earth. Heaven/paradise are shepherds. The Scripture says the meek will inherit the earth. Civilization began with the mining of gold! The miners were primitive men who hadn't built cities yet. The scientific discoveries verifying this go back hundreds of thousands of years. They were primitive. But not for long. It was the most precious and sought after element on earth. Sought after by everyone, for the gods! Why? We know that primitive man didn't need it. Why would 'spirit' genie gods need it?

"This should make us question our religious descriptions of them as genies. It should make us question, why gold? All that we need to do as scientists is trace the physical evidence. And then see if modern science can corroborate any scientific importance of it. The evidence of gold's ancient scientific use is found with the Sumerians. The evidence of today's scientific uses for gold lie with NASA and space travel. The Sumerians gave us the Jews. The Jews give us a scripture that supports why the Sumerian gods needed it: 'Heaven's streets are paved with gold.' The Sumerian gods needed it to repair the ozone layer of the tenth planet. We need it for ours. It has a big hole that is causing global warming.

"Is it possible that we need it for the terra forming of Mars? It is the sixth planet from the tenth. It is our closest planet and the most logical to inhabit. The Hopi petroglyph even shows this purification to be a removal of the man's majority to space.

"Notice the similarity of the transporting vehicle to a truck trailer." I put the drawing on the projector. I showed them the one shepherd left behind.

"At the end of this presentation I will make a case for our previous existence on Mars. This is one of my examples. But let me continue to make my point about our purpose for them, their motive. And why they don't just show up. If you remember earlier, I concluded that religion began in the absence of the gods. They didn't openly co-habitate with civilized man. Only with primitive man. I theorize the reason is that we were too powerful and brutal. Also, because of whom I theorize we are. Which I will get to briefly and use evidence to answer why they don't show. First, I want to give you my last chain of evidence to answer the motive for our creation. That is to mine gold. Gold is power. Money is the root of all evil. And money can buy all the people you want. Hell, we can even make them, now. All this is possible. And beauty of the flesh can get you all the money you want. Most men would use it to buy an island and stock it all with beautiful women. We don't like being ugly and we don't like ugliness. It limits our power. One affects the other and together they're a literal ménage à trois.

"Now, to finish the evidence verifying the aliens being the gods of primitive man and needing gold. The cave drawing of the wandjina aborigine god is alien-looking and has a halo around his head." I put the drawing on the projector. It is a drawing by primitive man. But not before I pointed out the big headed Hopi god again. "The symbol that civilized man gave for God/gods was the gold halo above the head. He made him man! Finally, the evidence to answer why they don't show."

"I will start by saying I think this answer involves the science of quantum physics and a new understanding of time travel. It is an extremely difficult subject matter, and the men behind it we call geniuses. Albert Einstein is the most famous of all. Stephen Hawking is the current leader in this field. His book is called *The Mystery of the Universe in a Nutshell*.

"Yeshua said heaven is like a mustard seed. Plant it, nurture it, and soon it is everywhere. They both strive to answer the mystery of the universe. They approach it by tracing it back to its beginnings. A seed! The physical evidence doesn't lie. Man lies! The quest for the beginning of the universe led Hawking and Einstein to the atom and the big-bang theory. The atom is invisible energy. We see it, but when we look closer at what makes it, we don't see it. The last thing we see is an e-string.

It is a tube of photons, information. Light! Any variation of it makes sound. This is what we think makes up everything and connects it. Einstein was working on a Unified Field Theory at his death. This basically says that everything is connected. The influence at the quantum level affects the outward result of the atom. There are three basic shapes: the circle, the straight line and the wave. This all can be seen on the computer, which certainly has made our informational world global or unified through the same principle—fiber optics! Our bodies resemble this framework, as well as the DNA structure of a single cell. I am not an expert, so some of this could be incorrect. However, I am trying brainstorming and using these examples, to answer the question of why they don't show.

"I theorize that time travel happens through the body and not in a machine. We are making an electrical image or motion picture of time as we exist. And when we see ghosts or other anomalies of time, we are seeing that electrical image, like a hologram. I theorize the aliens don't show because we are them, in a body that we shouldn't be. We are addicted to its power and must save or cure ourselves. We can't ever stop existing. Once we exist we always exist. We ultimately have to be satisfied with our original outward ugliness. All the spiritual teachings are inner. I based this theory on the scripture about the angels in the Book of Jude. I theorize that the angels are gods/aliens. In fact, they are called sons of gods in Genesis.

"Then Yeshua makes it clear that we are gods. John 10:34. The quote by Enoch makes my theory very clear. He was the seventh from Adam. That's pretty close to the beginning of the Jews for me. It is also the basis for the first book's title. *The Two Witnesses and the Religion Cover-up*. Enoch is not part of the Old Testament. He writes about the first six who came. They weren't supposed to. And obviously the sons of God in Genesis 6:4 weren't, either. Shortly after there was a flood to rid the earth of their wickedness. Enoch describes them as exceedingly white. It describes the grays. But his quote in Jude makes my theory simple to understand. It is in verse six. Coincidence? 'The angels which kept not their first estate, but left it for a strange flesh' is punished in chains and eternal darkness. I say that is the body and earth. They both create gravity like a chain.

"Our problem is described like an addiction. 'We are like dogs returning to vomit, pigs returning to mud.' And we are most like pigs, physically and biologically. My theory is that we are the gods/aliens, in bodies that we created. We have already surpassed our level of technology and conquered space. We engaged in the making of bodies. We made them as beautiful as we could and it became the biggest business in the universe. Then we could become them. I theorize that our original body must be in a death state. And we are its memory in this body. The strange flesh. That's why we must be born again of the spirit. It is common to every culture. And this could be why flesh and blood can't enter the kingdom of heaven, our first body. Only our memory can enter. And this must be spirit.

"Like uploading information. I liken the process to the movie *Total Recall*. However, they use a virtual reality machine. I theorize that the mystery and our delta brainwave length give us this ability. It is achieved through the delta brainwave length or death state. This explains the way! It is in every culture. It happens when we experience mental death.

"Yeshua explains that a seed must die to yield fruit. We are the seed. Returning is the fruit. The tree of life. He describes being born again like the wind. 'Not knowing from whence it cometh or where it goeth.' He describes that day to be like lightning striking from the east and landing in the west at the blink of an eye. He describes the Unified Field Theory of Albert Einstein. He describes our ability to see when were out of the body. It's like being everywhere, an out-of-body experience. You're just seeing and have no body.

"My brother had one and it scared the hell out of him. It scared him because he had no body and was moving away from his own. He was moving! This supports my theory about traveling back to our original body.

"He didn't make it back. If our motive that created the mystery of man is for power with beautiful bodies, then the angel story makes sense. They thought they could make a more powerful body than their own, but it got out of control. And since there is no blinking and thinking magic to make it correct, it has to be done with physical labor. And nobody likes to do that. I don't. If I could blink a piece of art I would." That got a few laughs. "When they saw the by-product of our beauty, though, I theorize that it must've consumed them. They wanted to be like that. They couldn't if they all looked the same and were ugly.

"Hence the Jewish angel story in which one-third wanted to be greater than God for power. But they knew the outcome. And they knew they couldn't stop it. They also knew the cure. Maintain a majority two-thirds. Keep the minority in isolation and running its course. Time won't heal man, but man could save himself through time. This all sounded like a self-destructive computer program to me. Especially in the present age of computer viruses. Open it and play hell in trying to stop it. Man was a literal Pandora's box. A virus that was a necessary evil to ensure its own survival. We needed gold for the terra forming of earths. We became their competitive power-seeking robots. We are essential to the perpetual survival of paradise, earths! Heaven's footstools. They couldn't prevent the outcome of their desire for power. Magic of that kind doesn't exist. That's religion's downfall. No one can heal or raise the dead with a blink of an eye. But there is magic in science. It is invisible and takes time. It's called quantum physics."

I use two quotes by Yeshua to set up the future scenario. One on my first book and another on this sequel "If your days had not been shortened; No flesh would be saved," and "Many are first that shall be last and many are last that shall be first."

"We must learn who we are. We must seek the FUTURE! I propose that these words of the Bible describe the future and the reality of our existence. First and most important, being sent here. They don't want to come. But they need gold and to save us from an animal-like evolutionary existence. This supports mankind being hell. Those born of women support the reincarnation theory and the theory of genetically creating bodies to inhabit them—those not born of women, but in a laboratory-like cloning. It also supports the Immaculate Conception as a possible alien abduction. This sheds light on more than one race being created. It makes sense of a possible ongoing process of alien abduction and cow mutilation. This would serve the purpose for research being in the most lucrative business of the world. Human Creation!

" 'Reaping what we sow', 'Plucking up the trees of man that my father did not plant', 'Those born of women', 'I was sent to my father's children, the chosen race of the Jews'—he even called the Samaritan woman a dog; 'John didn't come into the world eating and drinking,' 'It isn't possible to be able to deceive the elect,' and last but not least, 'My father why have you forsaken me?'

"This makes it clear that death of the mind is the way. And we are the bride and our alien body is the bridegroom. Holy Ghost and spirit, the spiritual union of becoming one. It is all about science! There are no gods/aliens that are genies. If there were, then we must ask ourselves why it took three days to raise Yeshua. Why not right then? Why not now? Why six days to create man? And for goodness sake—yes, goodness—why would a loving father kill any of his children?

"I beg you all to open your minds to these possibilities. We should seek knowledge of our universe. Knowledge is power. The golden rule should be our guide. The love of our children should be the purpose. We should sacrifice ourselves for them or not have them. And in doing so, we will see the endless possibilities that are before us. Knowledge is power!

"However, it should become very clear that we could have existed before. And the knowledge we strive for has become life threatening, even for the planet. My last piece of evidence to support this possibility is on the last page. It is what threatens our existence! Our quest for power has given us knowledge that is now our inevitable downfall. A nuclear global war! We can't save us, only ourselves. This is my prediction of the future. We will experience contact, like Yeshua at his end, when we are ready to destroy the earth. This is written in Chapter 12 of both Daniel and Revelation. Contact is made to save the earth. And the answer to the mystery of life is known by everyone. I am not a religious fanatic, but this overwhelmed me.

"This is what should happen in my theory and scenario of man with the aliens. They are more advanced than we are. And when contact is made, we will immediately know the answer to this mystery. It's obvious that time is conquerable.

The Bible is mostly composed of prophecy, which is to predict the future. They will reveal themselves and disarm us. We will meet ourselves, and our ignorance will reveal us. If we haven't learned about ourselves, pride will make us fall. We must cure our addiction to outward beauty and power. We are outwardly ugly! I know that's relative, but I theorize we are innately addicted to outward beauty. We created it like a Doctor Jekyl and Mister Hyde. And then beauty became more than relative. Mankind became our evil twin."

I wanted to answer the question of how the universe created us, the aliens. I brainstormed one more time. "We look like an embryo, sperm itself. Could we be an internal being? One that never lived outwardly? We are bald, big headed, and have little muscle development. I can only guess. But I believe this kind of speculation will make way for scientific discoveries that could lead us to this answer.

"Finally, I will answer the question of the world. Could I be Michael, and who are the Two Witnesses? I have already stated my prediction of when this answer will occur, 2012. I theorize the five races of man will be at global war at this time. They will be threatening the very existence of all life on this planet. I will be ushered onto the world scene in New York City. I will speak at the Olympics and beg the leaders of the world to resist war. My death will happen before the eyes of the world."

Little did I know that Jake comes out of hiding in New York City. I would sacrifice myself. But I couldn't force Jake to do the same. I couldn't believe he showed up.

I went on. "I am afraid, but I believe in life after death. Therefore I make my stand for peace with peace. The future is now. How can anyone who believes in life after death kill? That's why believing is the final test.

"Can we make China disarm? I don't think so. Then why should we make Iraq? Do we? We will continue to have war to the end. I base this on our president, who says he believes in life after death, and yet endorses war. So do the majority of leaders in the world. They all have armies. Will the last Biblical war be with China? I am surprised at the prophecies' numerical match to them. They do have an army of two hundred million, which is the number that is prophesized.

Who attacks the leaders of the world? This is the last war before the Two Witnesses and contact. It could be, if we try and make them disarm. It really doesn't matter who. It only matters to believe in life after death! Then we will say 'O death, where is your sting?'

"I am often asked by these religious people, like our president, if I would let someone kill my children. I always state that I would give my life for them, but not by aggression. I would try my best not to. I would take the beating or bullet first. I believe in life after death, and "Turn the other cheek" is my stronghold. They always say they would kill anybody who tried. I can only say to you that I believe in peace. I believe that Martin Luther King Junior made the world a better place for his children. I believe my brother did the same in his own way. He couldn't control his anger. I hope that I can do the same. I love him, Martin Luther King and all those men who can make such a sacrifice for their children."

I took no more questions and wept out as I left the stage. I turned at the last moment and cried out, "Please forgive us all, father, for we know not what we do! But soon we will all know and then what will we do? Peace!" I yelled as hard as I could, "Peace, is the only answer." I kept yelling as my bodyguards rushed me away.

When we got into the van, my mother began to cry uncontrollably. "Please don't cry," I said as I held and comforted her. "It's going to be all right." I didn't realize it, but I was stroking her hair.

She looked up at me and smiled. It was the most beautiful thing I've ever seen. "I know it will, and when you wake up," she said with tears streaming down her face, "I will be doing this for you."

I knew that she knew. And we both knew her tears were not only tears of joy. They were also tears of pain, from knowing we will all see the hell mankind's children experience. We will even see our own and what we inflicted. Nothing is hidden that won't be revealed.

We all cried!

As we sped along, the driver suddenly turned on the radio. The United States attacked Iraq.

"No!" I screamed. "No!"

I blacked out. My nephew was there. I woke up and Mom was stroking my hair. But I wasn't in the van. I was lying on a bed in a, a... "What! Where am I?"

Mom was squalling. She was in a weird-looking suit. This time though, they were tears of joy. Why was she wearing this suit? A suit that even covered her head. Oh my God. There must've been an attack. "Have we been attacked? Mom, please tell me what is going on! Where are the kids?" I couldn't remember a thing. I was panicking.

I looked around. We were in a makeshift hospital on the street of a ravaged city. It was also enclosed in plastic. There were cameras there and cameramen were also in these suits.

"Mom, have we been nuked? Is this Jerusalem?" I cried.

"Yes. Yes it is," she cried back. "You wanted to be brought here, so we did."

"Was it bombed?"

"Yes, Michael. It was bombed right after you were," she hesitated.

"Killed," I responded weakly. "Is that what your telling me? That all this has happened, just like I wrote it?"

"No, Michael. You weren't killed in New York."

"Where was I killed, then?" I couldn't remember. I still thought I was back in 2003. Back in Cookeville, Tennessee, at my first book signing.

She hugged me as tight as she could. "Don't you remember what happened in New York City?"

"Mom, I can't remember past my first book signing in Cookeville."

"Well, your book became a huge success. You went to the Olympics in New York and challenged the Israeli scientist to debate his evidence. Then you were shot at, but it didn't kill you. The Israeli scientist refused to come to the United States. You and Jake came to Jerusalem for a worldwide debate over his alien mother goddess statue."

"Then how did I get killed?"

"You got killed at the press conference outside the hotel here in Jerusalem. And now thank God you're alive."

"No," I said quickly. I looked directly into the camera. "Thank the aliens!" I held up the universal peace sign.

I turned back to Mom and asked her where Jake was. I realized that I hadn't seen him.

"He's over there," she said, brimming from ear to ear.

I looked and he was doing mystery. "Jake, did you get killed, too?" I asked quickly. Then I saw his wound. I saw mine, too.

He opened his eyes and said, "It's time."

Then it happened. Resurrection, contact, and finally, peace!

I knew all along that he would be with me at the end. He's the only one that would do mystery. They wanted us to do mystery. It is everything because it is nothing. Time to upload. This time it ain't going to be pretty. They needed our gold, and we need them. We are them. They established peace. WWYD.

Look forward to the final book, **The Future!**

MICHAEL

www.ingramcontent.com/pod-product-compliance
Lightning Source LLC
Chambersburg PA
CBHW021059210326
41598CB00016B/1256